First published by the Natural History Museum, Cromwell Road, London SW7 5BD.
© The Trustees of the Natural History Museum, London 2020

Published in North America, South America, Central America, and the Caribbean
by Smithsonian Books

This book may be purchased for educational, business, or sales promotional use.
For information, please write: Special Markets Department, Smithsonian Books,
P.O. Box 37012, MRC 513, Washington, DC 20013

ISBN 978-1-58834-671-1

Library of Congress Cataloging-in-Publication Data
Names: Kenrick, Paul, author.
Title: A history of plants in fifty fossils / Paul Kenrick.
Description: Washington, DC : Smithsonian Books, 2020. I Includes index. I
    Summary: "An illustrated history of plants presented through the stories
    of 50 key fossil discoveries"-- Provided by publisher.
Identifiers: LCCN 2019049623 I ISBN 9781588346711 (hardcover)
Subjects: LCSH: Plants--Evolution. I Plants, Fossil.
Classification: LCC QK980 .K43 2020 I DDC 580--dc23
LC record available at https://lccn.loc.gov/2019049623

Printed in China, not at government expense
24 23 22 21 20                       1 2 3 4 5


Internal design by Mercer Design, London
Reproduction by Saxon Digital Services
Printed by Toppan Leefung Limited

Front cover: *Acer trilobatum*, maple leaf, found in Ohningen, Baden-
Württemberg, Germany. This fossil dates back to the Miocene period and has a
width of 3 in (9 cm).

# Contents

# Introduction

It's not easy being a plant. You are rooted to one spot where you remain for the duration of your days. This means that all nourishment must come to you, and your immobility makes you vulnerable. You will never be able to run away when things take a turn for the worse. Even when ideally situated, you are unable to seek out and meet your partners in life. You wait for pollen to arrive on the breeze or as little packages brought to you by animal go-betweens. It is these latter that you aim to attract, and you do so with a flash of colour, a whiff of scent and a sugary treat. You abandon your numerous progeny to their fates at an early stage in their lives. They are cast into the wind or wrapped up in tasty packages for unwitting animals to carry off. Hopefully, of those that are eaten a few will survive. You are home to all sorts of animals that are difficult to shake off, particularly insects. Many of these are pests that only want to nibble at your leaves or tap into the juices inside your tissues. Then there are the fungi. Some of these are actually friends – well partners at least. Fungi are good at scavenging nutrients from rocks and soils, so you encourage them to take up residence in your roots, rewarding their labours with sugar. You try to fend off other harmful fungi. As for humans, no other animal makes use of so many of your kind. The relationship is exploitative at best and downright careless and destructive at worst. You are used in buildings, furnishings, clothes and as foods and medicines, but also you have broader cultural influence that encompasses gardening, the arts and the spiritual. You are present everywhere, yet we humans know so little about your kind and your ways. You are taken for granted, yet when we uproot or cut down too many of you, we diminish the quality of our lives and threaten our very existence. With the possible exception of bacteria, no group of organisms has so profoundly affected the long course of life on Earth. Plants have existed for more than a billion years, quietly going about their business, both influential and resilient. In this book we dig into the rocks, unearthing the plants within to help tell this little-known story. What emerges are forms that are, at first, puzzling yet frequently striking and beautiful. Upon careful investigation these give up their botanical secrets leading us along paths into lost worlds.

Though animals tend to grab all the attention, the breadth of the geological record of plants is quite extraordinary. Petrified tree trunks are the largest and heaviest fossils on Earth, whereas grains of pollen are among the tiniest, lightest and most abundant. Plants become fossils in a remarkable variety of ways. Commonly they fall into the muds and sands of lakes and rivers and, over time, are transformed into thin films of coal that outline the shapes of stems and leaves. Sometimes plants preserved this way are so abundant that they make vast seams of coal, which we have mined for generations to fuel our energy needs. Plants don't have a ridged skeleton, but the walls of the cells that make up their tissues are extremely robust. They are composed of cellulose, which in wood and also in pollen becomes encrusted in complex polymers making them stiff, repelling water and resisting decomposition. These properties mean that plant cells are commonly preserved in mummified remains and charcoal or where they have been infiltrated by minerals during the process of fossilization. Such fossils give us access to the internal tissue systems of ancient plants, providing insights into their evolution that are seldom available with fossil animals.

Plants naturally fragment into pieces, shedding vast quantities of leaves, seeds and pollen over their lifespans. Typically, it is these elements that become fossilized. So, unlike animals, an individual plant can give rise to many fossils, and these bits and pieces can be diagnostic of species, genus or family. Such an array of possibility presents the palaeobotanist with a bit of a puzzle. There are also false leads that have to be considered. For example, a few crystalline growth forms of minerals can resemble plants, as can some colony-forming marine invertebrates, while bones are sometimes confused with fossil wood. Because of the disassociation of parts, the whole plant can be difficult to visualize. Piecing large plants together is often not possible or may take the labours of many people over years, relying upon serendipitous finds that provide clues to long-lost associations. When this happens it can lead to unanticipated discoveries of marvellous extinct forms unknown in the modern world. This is part of the challenge and excitement of the field.

Plants originated in the world's oceans, moving from there onto land via rivers and lakes over 500 million years ago. This world-changing event led to life as we know it today. This book is divided into seven loose themes that explore the evolution of plants and their broader influence. The first plants to make the transition to land were both simple and tiny. This ancestor, which acquired the capacity for photosynthesis from the bacteria, probably took the form of an inconspicuous green alga that grew as filaments or small packets of cells in damp soils (origins theme). From this modest and unpromising foundation, plants underwent one of the most spectacular evolutionary

BELOW: A false fossil, the manganese oxide mineral dendritic pyrolusite that resembles the leafy branches of a plant, from the Late Jurassic 150 million years ago, Solnhofen, Bavaria, Germany.

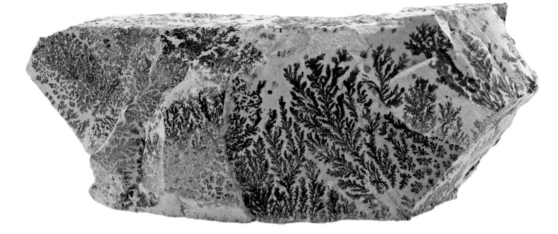

developments ever seen in life on Earth. They first evolved basic organ and reproductive systems (seeds, roots and leaves theme) and later fruits and flowers (flowers theme). Countless new forms arose (ancient plants theme) and their size increased by several orders of magnitude giving rise to the trees and shrubs that populated the first forests (rise and march of forests theme). The landscape that the first plants colonized was teeming with bacteria, fungi and protists that inhabited shallow soil crusts. Within these varied communities of microbes alliances were forged, leading to symbiotic relationships, notably between plants and fungi. Later, plants evolved myriad tricks to entice animals to their cause along with an armoury of weapons – many of them chemical – to fend off foes. These testify to the importance of co-evolution, in which a change in one organism elicits a response in another and so forth (friends and foes; flowers themes). Fruits and flowers are two of the most obvious consequences of the long-established relationships between plants and animals. Plants are hugely influential to us, which is touched on in the final theme. Organisms and their relationships are one side of the story, but plants interact with the physical environment, too. Over millions of years there has been a dynamic feedback between plants and large-scale planetary processes (climate and diversity theme). As the cloak of green crept across the surface of the land it interacted with atmosphere above and the rocks below invigorating the cycle of nutrients, altering the chemistry of air and water, and gradually influencing the climate of our entire planet. Such large-scale interactions are not evident in our everyday lives, they make themselves felt over millennia, but they are crucial to our well-being and to the long-term habitability of Earth.

# The first photosynthesis
## banded iron

Undulating bands of orange and red snake their way across the cut surface of this 2.6-billion-year-old rock. The striking colours are due to the iron oxides contained within. These are known as banded iron formations, which are found in rocks of broadly similar age worldwide. This one comes from the Hamersley Basin in Western Australia, one of the world's most important sources of iron ore. The deposits formed in shallow seas as iron in solution was oxidized to less soluble forms either directly or indirectly by photosynthetic organisms. Between about 2.4 and 2.1 billion years ago there is further widespread evidence of the rusting of iron in soils and the disappearance of other easily oxidized minerals from ancient streambeds. These notable chemical changes in the rock record mark the first appearance of appreciable quantities of free oxygen in the atmosphere. So noticeable is this that geologists call it the Great Oxidation Event, although recent research shows that it was more of a protracted and dynamic transition, a sort of rollercoaster ride, as rising levels of free oxygen produced by photosynthesis were drawn out of the atmosphere through the oxidation of vast quantities of minerals and gasses near the Earth's surface. Eventually photosynthesis won out, leading to the rise of oxygen and the greatest environmental transition in Earth history.

Today, oxygen makes up 21% of the atmosphere, but vanishingly small amounts of this gas were present during the first half of Earth's 4.5-billion-year history. The story of oxygen begins with the evolution of photosynthesis in bacteria. Photosynthesis is where an organism converts sunlight into chemical energy and there are several ways that bacteria do this. Crucially one type of photosynthesis had a by-product of oxygen. This oxygenic form arose in a group of bacteria called the cyanobacteria as long ago as 3 billion years. In this type of photosynthesis energy is generated from sunlight, carbon dioxide gas and water. The oxygen thus produced is a highly reactive gas that leaves its fingerprint in sediments, particularly in the way that it oxidizes minerals. When these readily available oxygen sinks became saturated, the gas began to accumulate in the air. How long this took and how it enriched the atmosphere is still much debated, but oxygen concentrations varied between 1% and 10% of present levels about 1.5 billion years ago.

The rise of oxygen had profound and far-reaching consequences. It changed the chemistry of our oceans, reacting with dissolved iron causing it to precipitate out of solution. At the same time, the oxidation of minerals on land altered the proportions of elements that weathered from rocks and travelled through rivers to the oceans. One consequence was that our oceans became enriched in sulphates, as they are today. Oxygen is toxic to anaerobic organisms, so these once dominant life forms were forced to the margins, finding refuge in places where oxygen could not penetrate, such as deep within sediments. As greenhouse gasses like methane, which was abundant in Earth's early atmosphere, became oxidized, Earth's climate cooled contributing to a worldwide glaciation known as Snowball Earth. Against this backdrop of environmental change, life began to evolve from single cells or simple filaments giving rise to a great variety of complex multicellular forms. The evolution of the rich diversity of organisms that followed, including animals, was dependent on free oxygen, which is crucial to the metabolic process of respiration. In a further remarkable transformation, this ancient mechanism of oxygenic photosynthesis was later incorporated into other organisms, some of which evolved into algae of several different kinds, including those from which plants evolved. Our history of plants therefore begins with the origin of oxygenic photosynthesis in bacteria long before the first plant had evolved.

# Plants evolve
## tip of red alga

A narrow filament made of stacked discs topped by a wider zone composed of many brick-like cells forms the tip of this 1.2-billion-year-old red alga from Somerset Island in the Canadian Arctic. Individual whole plants were tiny – barely visible to the naked eye. The hair-like filaments branched infrequently and were anchored to the substrate by a special clasping holdfast. In life, one can imagine these growing as purple-black to rust-coloured clumps forming a felt-like turf that encrusted patches of ground in the pale, limey mud at the shallow margins of a lagoon. This tiny fossil claims many firsts. It is a eukaryote, meaning that unlike bacteria its cells contained a membrane-bound nucleus and mitochondria. It is the oldest eukaryote that can be classified with confidence into a known major group and the first that was unambiguously photosynthetic. Although these attributes are not directly visible in the fossil, their presence can be inferred because of its many other similarities to the modern red alga *Bangia*. Structures observed in the fossil also indicate that it had a sexual life-cycle, again by comparison to modern *Bangia*, making this the earliest unequivocal evidence of sexual reproduction. Developmentally, the fossil stands out as distinct from prokaryotes in other ways too. Various cells took on different shapes, ranging from disc-shaped to wedge-shaped; other cells took up special functions, becoming anchoring holdfasts to the upright filaments. This cellular differentiation is early evidence of one of the trademark developmental mechanisms of eukaryotes, in which cells proliferate and become specialized for specific functions, eventually evolving into tissues and organs, leading ultimately to larger more complex organisms.

All plants, including red algae, green algae and the more familiar species growing in our fields and gardens, acquired the capacity for photosynthesis in the same way – directly from cyanobacteria. The ancestor of plants was a predatory eukaryote that engulfed a free-living cyanobacterium. Instead of being digested, somehow the cyanobacterium survived and was retained inside the cell, henceforth to be passed on to following generations. Over time, the captured cyanobacterium was domesticated, becoming so integrated into the machinery of the cell that it lost its ability to live independently. This theory of the origin of the chloroplast has gained overwhelming support from the study of plant and cyanobacterial cells and their DNA. The acquisition of photosynthesis and the evolution of more complex tissues and organs in eukaryotes over 1.2 billion years ago marked a radical change from a world dominated by simple prokaryotes to one in which living organisms became ever larger and more varied.

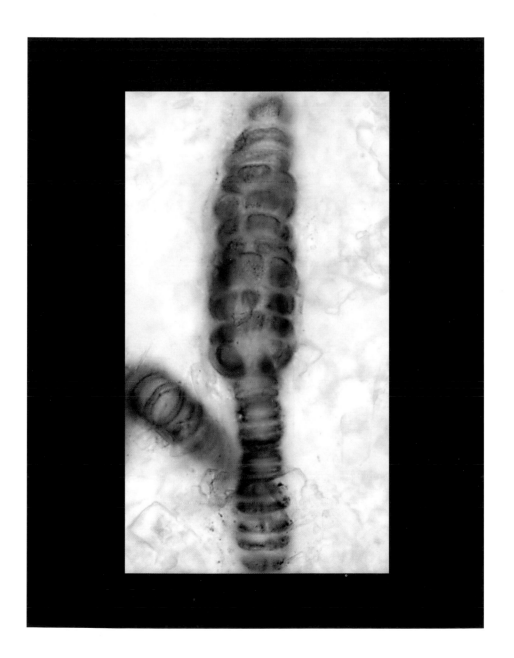

# Algae flourish
## *Coelosphaeridium*

For many years geologists puzzled over the identity of these curiously shaped fruit-like balls with their inverted lining of cones. About the size of peas and called *Coelosphaeridium*, meaning hollow spheres, these 465-million-year-old fossils were variously interpreted as shells of amoeboid protists, sponges, bryozoans, early relatives of starfish or sea urchins, even the eggs of large snails. Today they are known to be green algae, belonging to the Order Dasycladales, where they are part of an extinct tribe whose precise relations are still debated. Modern species live mainly in sheltered bays and lagoons in shallow tropical seas. Unusually, individuals are entirely constructed of one enormous cell. Most unicellular organisms are microscopic, but Dasycladales are an exception, being clearly visible to the naked eye. Deposits of calcium carbonate coat the surface into which the external features of the cell become engraved. It is these coatings of lime that are preserved as fossils, and under certain circumstances they are so abundant that they can actually form the bulk of the rock itself.

Today, algae of various kinds are the most important primary producers in the oceans, synthesizing organic compounds from carbon dioxide and water with energy obtained directly from sunlight. They are very varied in form and include red, green and brown macroalgae as well as vast numbers of microscopic plankton. We have seen how the red algae, the green algae and plants all acquired photosynthesis directly from an engulfed cyanobacterium, which evolved into the subcellular organelle called the chloroplast (see p. 10). Many lines of evidence including biochemistry, cell structure and, more recently, the genome itself, show that the evolution of photosynthesis in other algae, such as the brown and yellow-green algae, took different routes.

Brown algae are important seaweeds of the colder waters of the northern hemisphere, where some species of kelp can exceed 60 m (197 ft) in length. The type of chloroplast found within their cells was originally derived from a unicellular red alga. A whole red algal cell was engulfed and retained within a cell of the ancestor of brown algae, where it became the site of photosynthesis. Over time, the original red algal cell lost much of its contents, apart from its photosynthesizing chloroplast. Thus, the chloroplast of brown algae is a product of two independent acquisitions of photosynthetic cells on separate occasions followed by domestication of these cells to create the subcellular platform for photosynthesis. The large brown algae of our seashores are not close relatives of plants. Evidence from their DNA suggests that they share an ancient common ancestor with diatoms and water moulds.

Many of the microscopic algae that now dominate the phytoplankton of the world's oceans also occupy remote branches of the tree of life, and likewise they acquired their capacity for photosynthesis mainly from the red algae. These include the coccolithophores, which are spherical cells that bear shield-like calcareous plates, and the ubiquitous diatoms, which have cells of varied shapes that are enclosed by a distinctively patterned box made of silica. Although tiny, the cells of dead coccolithophores and diatoms are so abundant that they form muds rich in lime and silica in modern ocean sediments. In the geological record, chalk rocks are mainly composed of coccolithophores, while diatoms are the main ingredients of deposits of diatomaceous earth. Another group of unicellular phytoplankton called the dinoflagellates can form blooms that colour the water, colloquially known as red tides. Other species are bioluminescent, emitting a blue-green light. About half of the species of dinoflagellates are photosynthetic, mostly also following a red algal route to photosynthesis, similar to that taken by brown algae. The breathtaking diversity of plant-like organisms in the oceans reflects the independent acquisition of photosynthesis by aquatic eukaryotes on several far-flung branches of the tree of life.

The history of algal evolution in the oceans is most conspicuous in the fossil record of macroalgae, particularly in those encrusted with lime, of which *Coelosphaeridium* is among the earliest. Other clues come from minute yet abundant fossils of the cyst-forming stages of the phytoplankton and from the geochemistry of hydrocarbon residues sealed inside rocks. Algae and cyanobacteria are important contributors to the world's crude oil reservoirs where they leave distinctive chemical fingerprints. Evidence from hydrocarbons and fossils shows that the primary producers of the oceans have become more diverse over time. Cyanobacteria were the first to dominate the oceans of the Proterozoic Eon. Near the onset of the Phanerozoic Eon, about 550 million years ago, they were joined by green algal phytoplankton and later by green and red macroalgae, which became ecologically important as part of life on the sea floor. In the modern world, the diatoms, dinoflagellates and coccolithophores predominate in nutrient-rich shallow seas leaving the nutrient-poor surface waters of the mid-ocean to minute cyanobacteria and green algae. Brown algae are also latecomers to temperate shores, beginning to diversify during the Cretaceous Period. This huge variety of primary producers in the oceans stands in marked contrast to the situation on land, in which photosynthesis is dominated by a single lineage of plants, all of which trace their ancestry to the green algae.

# First land plants
## *Cooksonia pertoni*

During the 1920s the botanist William Henry Lang became intrigued by some puzzling fossils in ancient rocks. While inspecting 420 million year old sandstones in a roadside quarry, he noticed that they contained thin, irregularly shaped, films of carbon. From among the background clutter of organic fragments, he observed a tiny forked strand with trumpet-shaped ends. About half the size of a dressmaker's pin, this fossil was incomplete, its lower parts being indistinct or missing. From these unpromising remains, Lang was able to extract spores and some internal tissues, establishing beyond doubt that this was a plant that once grew on land. He named it *Cooksonia pertoni* in honour of the Australian botanist Isabel Cookson and in recognition of the place in which it was discovered, the hamlet of Perton in the English county of Herefordshire. Since Lang's discovery many species of *Cooksonia* have been found in rocks of similar age around the world, and this diminutive plant has become famous as the first plant that we know of that was truly capable of living on land.

Modern analogues of *Cooksonia* are hard to find, but in its simplicity and mode of function it resembles the tiny spore-containing capsules of mosses, which are borne on the ends of long, narrow, unbranched stalks. In mosses, the spore capsule and its stalk develop from a plant with leafy green stems. Seemingly one plant, these are in fact two separate stages of the life-cycle. The leafy plant is long-lived and it produces the sexual organs. The spore capsules are the products of sexual reproduction. They are short-lived and often develop on a seasonal basis. Raised up above the ground, the moss capsule can launch its contents of minute spores into currents of air, enabling them to be dispersed. *Cooksonia* is undoubtedly a spore-bearing plant, but it is not a moss. The sexual phase of its life-cycle still remains a mystery.

Fossils from other slightly younger geological sites shed further light on the nature of early communities that lived on land. The largest and most varied land-dwelling animals were the arthropods. In the wetter parts of the landscape, close to or submerged in shallow pools, early relatives of the springtails and the fairy shrimps lived alongside and perhaps fed on green algae, fungi and other protists. In the drier parts of the landscape, truly terrestrial species could be seen, including tiny mites and larger predatory relatives of the centipedes and spiders. Other animals included free-living nematodes, which, like their modern relatives, probably fed on bacteria. This fauna was important to the development of soil

communities. Alongside the tiny plants that made up the flora were abundant mat-forming cyanobacteria, which played an important role in the nitrogen cycle. Much like today, inorganic nitrate compounds, essential for plant growth, were scarce. Unlike plants, some cyanobacteria can fix nitrogen themselves, taking the gas from the atmosphere and turning it into a form that can be used by other organisms. Fungi were also abundant, playing important roles in decomposition and the recycling of carbon, as they do today. The fruiting body or sporocarp of one fungus could reach such great sizes that, when first discovered, it was thought to be the trunk of a tree and named *Prototaxites*. Fungal sporocarps were the largest living structures on the landscape, greatly exceeding plants in both height and bulk.

Fossils from these early communities provide evidence of interspecific co-operation, in which two different organisms interact to the benefit of each other. The earliest examples come from thin, layered crusts made of filaments that enveloped and harboured minute spherical cells. These crusts are now known to be composite organisms resembling living lichens. They were made from a fungus and a photobiont, which is either a green alga or cyanobacterium. In lichens, the fungus benefits from the carbohydrates produced by the photobiont, which in turn benefits from the sheltered microenvironment afforded by the fungal filaments and by the capacity of the fungus to scavenge nutrients from the environment. The closest modern analogues of these early communities are thought to be biological soil crusts and lichen and bryophyte communities that, though often overlooked today, are widespread and resilient. They can be found in ice-free regions of the Antarctic Peninsula, carpeting vast swathes of tundra and boreal forest, thriving in hot deserts, and encrusting the stonework in urban and rural environments. These simple pioneer communities created the conditions in which more varied life on land could flourish.

Fossils of *Cooksonia* tell us that the earliest plants to grow on land were both very tiny and simply constructed; they were so small and simple that they possessed few of the features that we associate with plants today, like leaves, roots and seeds. In contrast, the animals that made the transition to life on land began their journeys with most of their fundamental tissues and organs already in place, including limbs, eyes and nervous systems that evolved in their aquatic forerunners. So, plant life began simply, and the varied forms that we see today took shape during a long history of interaction with soil and atmosphere, the latter shaping the evolution of plants perhaps more than any other group of organisms. Although firmly rooted in soil, plants are truly creatures of the air.

# The rise of oxygen
## charcoal xylem

In this fossil, the hole is the central channel through an individual xylem cell in the vascular system of an early plant. The walls have an inner lining of thickenings arranged in successive rings and sometimes in a helix. To the outside of each cell is a narrower wall with perforations. The detail in the structure of the cell wall of this 415-million-year-old fossil is remarkable, and is due to the original plant stem being charred by fire and turned into charcoal. As charcoal forms, much of the plant's original tissues are converted into compounds composed of only carbon and hydrogen that are organized into multiple ring-shaped molecules. Inert and brittle, these hydrocarbons are highly resistant to decay and physical compression, preserving in minute detail the structure of the cell wall. Fossils of this type provide insights into the evolution of plant cells. They are also among the earliest geological evidence for the phenomenon of wildfire.

Fire plays a significant role in life on Earth and has had a pronounced evolutionary effect on the flora and fauna of most ecosystems. Vegetation is particularly fire-prone in moist climates that have extended hot and dry seasons. Moisture promotes plant growth providing a ready source of fuel that will ignite when dry. Wildfire also has important effects on a global scale, regulating the concentration of oxygen gas in the atmosphere. Under natural conditions, fire is often ignited by lightning strike, but oxygen is also crucial to its ignition and spread. Today, the concentration of oxygen in the atmosphere is 21%. If this fell below about 15%, fires would not ignite, and below about 17% they would not spread. The consistent presence of charcoal in the sedimentary record indicates that oxygen in the atmosphere must have exceeded a minimum threshold of about 15% continuously for the past 420 million years.

The colonization of the land by plants is thought to have been responsible for the rise of oxygen to present atmospheric levels, adding to the oxygen already produced by marine algae. Although respiration ultimately consumes almost all oxygen produced by photosynthesis, returning carbon dioxide gas to the atmosphere, a small fraction of organic carbon escapes oxidation through burial in sediments. This is one way in which oxygen gradually accumulates in the atmosphere over geological time. The controls regulating this process are complex, but greater plant biomass on land tends to lead to more burial, which tends to add to the oxygen reservoir, whereas phenomena that reduce plant biomass, such as wildfire, act as a negative feedback. Delicately balanced natural regulators acting together have kept our atmosphere well oxygenated over hundreds of millions of years.

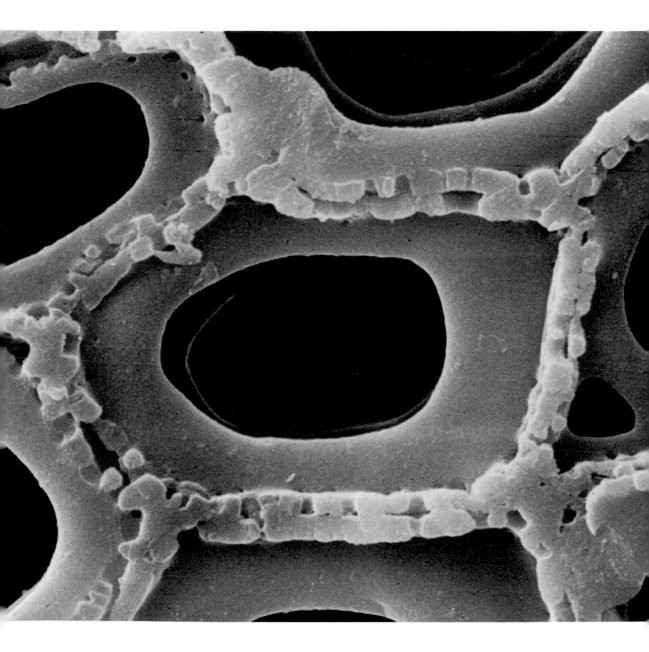

# Leafless plants
## *Thursophyton elberfeldense*

A tangled mass of stems captures the essence of early plant life on land, which was both bountiful and untamed. The 400 million year old rocks of the lower part of the Devonian Period give us our first really good glimpse of this early vegetation. The sediments entombing these fossils formed in a river or lake so they preserved plants that grew in close proximity to the water's edge. The plants were small, mostly less than 20 cm (8 in) tall, and they had simple forked stems. One striking feature is that most of the stems are smooth, but some sport a fine hair-like fringe. There are no leaves of any kind. In fact, fossils such as these tell us that leaves and roots had not yet evolved. So, how did these plants manage to thrive on land? Some indications can be gleaned from sections of stem that became infilled by minerals. These reveal microscopic details showing that internally they possessed the core tissue systems necessary to sustain plants of this size on land. The stems that are now brown were once green supporting photosynthesis, and in life they were anchored to the ground by tufts of minute cells called rhizoids. There are similarities here with modern liverworts and mosses. Why some plants possessed a fringe of fine hairs and others did not is still a puzzle. Perhaps these in some way facilitated photosynthesis. Another possibility is that they were a deterrent to small arthropods intent on piercing the stems to feed on sugars contained within. From unpromising beginnings in plants like these, leaves and roots evolved gradually, taking on their own particular characteristics in different plant lineages. One theory postulates that the evolution of leaves began in response to changes in Earth's atmosphere. At this time, levels of the atmospheric gas carbon dioxide were exceptionally high, and it has been suggested that this formed a barrier to leaf evolution. As levels of this gas gradually fell, a physiological threshold was eventually passed in which plants were able to sustain leaves with broad blades. With the environmental barrier to evolving leaves removed, a more efficient platform for photosynthesis ensued and leaves developed in various forms.

In a twist to the story of the evolution of leaves, plants are thought to be the main driver behind the falling levels of atmospheric carbon dioxide gas that characterized this early period of life on land. As plants first began to colonize the land surface they increased both the weathering of rocks and the burial of the organic carbon they produced through photosynthesis. The net effect was to lock up carbon dioxide gas in solid form in rocks and sediments. Thus, over many millions of years, plants affected the chemistry of Earth's atmosphere, which in turn shaped their own evolution.

# Plants take to the air
## spores in coal

Golden brown and scaly in appearance, these are the massed remains of spores, which are minute dispersal capsules that are produced in prodigious quantities by mosses, ferns and similar plants. When these spores became fossilized they were in such abundance that they make up almost all of this ancient Russian coal from deposits near Moscow . Each spore was originally spherical, but it has been flattened by the weight of overlying sediment into a saucer shape. Each one measures less than half a millimetre in diameter and no more than a twentieth of that in thickness. One cubic metre of this coal might therefore hold a mind-boggling 160 billion spores.

This coal is a lignite, meaning that changes to the original plant litter are slight; unusually for coals of this great geological age, it has not been subject to intense post-depositional pressure and heat. It formed during the Carboniferous Period in tropical equatorial regions when the continents were in very different positions than today. The spores that make up the coal came from long extinct relatives of modern clubmosses and quillworts, which nowadays are rather small herbaceous plants that live on the forest floor, in the canopy as epiphytes, or in lakes as submerged aquatics. The forebears of these modern species grew into tall trees that had distinctive pole-like trunks topped by a crown of forked branches creating ancient coal-swamp forests with a high, diffuse canopy (see pp. 39 and 63). Many species reproduced only once during their lifespan. It is thought that they grew rapidly to maturity over 10–15 years, developed one flush of cones that delivered a short-lived but massive output of spores, and then died. This mode of reproduction might explain the enormous quantities of spores that are found in some coals of this age.

Spores are produced in the life-cycles of seedless plants grouped in the bryophytes, clubmosses and ferns. Typically, they are airborne. To function in this manner they need to be light and resilient. Most spores are therefore tiny and invisible to the naked eye. En masse, however, they can take the form of pale yellow clouds or layers of dust. Each spore is equipped with a tough coat made of sporopollenin, which is one of the most stable and chemically inert of all biological polymers. This protects the delicate cell inside from drying out and from physical damage. On arriving at a suitable site, spores germinate, take root and develop into the sexual phase of the life-cycle. Fertilization is reliant on the presence of films of surface water through which minute sperm swim to reach the egg cell. Once fertilization has taken place, a new spore-producing plant

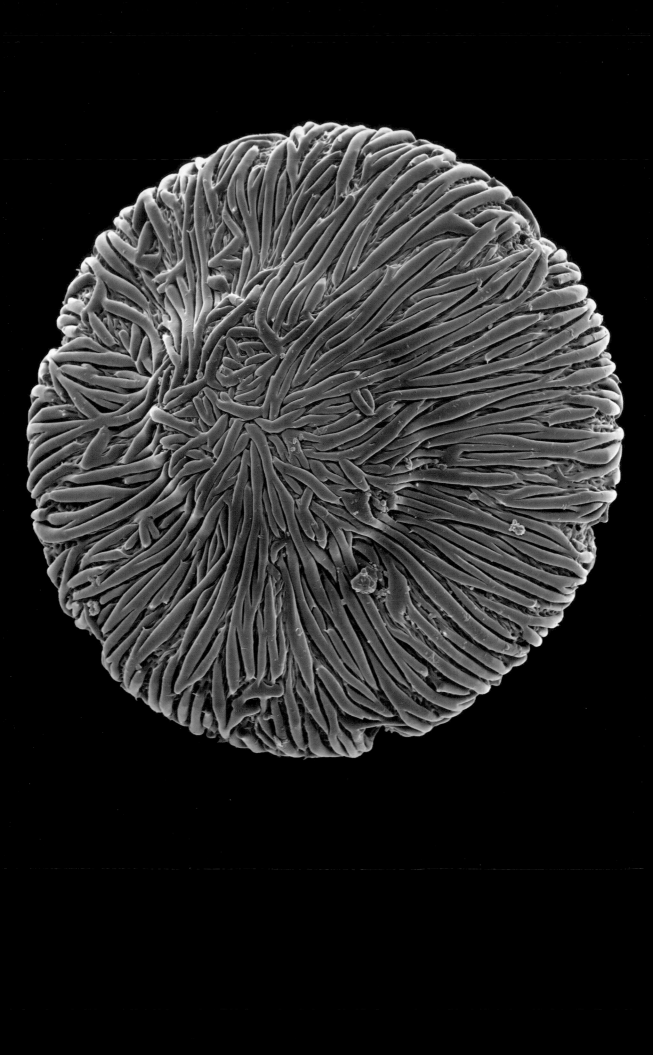

OPPOSITE PAGE: Scanning electron micrograph of the pollen of *Browallia speciosa*. Surface features of pollen are like fingerprints that enable us to track the distribution of plants through time.

develops. Spores are thus both an integral part of the plant life-cycle as well as an effective means of dispersal. These properties enabled the first land-dwelling plants to expand their ranges, moving from wetlands associated with river systems into the dryer hinterland and eventually to cross continents.

Other plants have modified this life-cycle through the adoption of seeds. These include the flowering plants, conifers and their kin. In these groups pollen travels from one plant to fertilize another. Pollen grains transport the male gamete to the female ovule that is enclosed within a flower or cone. Once the ovule is fertilized, it becomes a seed. The transfer of pollen can be mediated by wind, water or by animal vectors including beetles, flies, bees, moths and birds. Wind-borne pollen typically is small, very lightweight and produced in great quantities. Animal-pollinated plants produce pollen that is relatively heavy, sticky and protein-rich. Most pollen has a tough outer wall and this can be elaborately ornamented with miniscule spines, pores, warts and ribs. These distinctive surface features play roles in the transport, capture and germination of the pollen grains. The seeds formed in this way now take on the dispersal function, which is akin to the role of spores. Like spores, seeds have tough outer coats, can travel far from the parent plant, and have the ability to lie dormant until favourable conditions arise for germination.

For palaeobotanists, pollen and spores hold a further secret that makes them a useful and faithful recorder of the history of plant life. In many the intricate surface features can be so distinctive as to provide a sort of fingerprint that enables the dispersed grain to be traced back to a particular group or species of plant. Magnified over 3,200 times using an electron microscope, the pollen of *Browallia speciosa* from the nightshade family Solanaceae reveals a characteristic meshwork of grooves. It is the study of details like these that enable us to attribute the Russian spores to ancestors of the clubmosses and the quillworts. The resilient spore wall that serves the plant so well during dispersal becomes, after death, an enduring testimony to the former existence of a species. Spores have been captured and preserved in sedimentary rocks ever since plants first began to colonize the land. By virtue of their unimaginably vast numbers and their capacity to travel long distances, they are more abundant and found in a wider range of rocks than are the plants and plant organs that gave rise to them. Spores and pollen therefore often furnish the earliest fossil evidence for events in plant evolution, and changes in their relative abundance through sequences of rock are our best indicator of the shifting patterns of vegetation through time.

# Seed evolution
## *Xenotheca devonica*

In the English county of Devon, on a windswept promontory facing the Celtic Sea, sits a small roadside quarry that yields the fossilized remains of an ancient shrub. The fossils are difficult to extract from their thin bed of sandstone, and during the process mostly break into pieces a few centimetres in length. When more complete, they show a plant that was shrubby, with narrow, wiry stems and leaves. Their rusty autumnal hue is not original but reflects the presence of iron oxides in the sediments in which the plant was buried. The tiny claw-like branches pointing skywards mark a fundamental change in the way that plants reproduced, and they were harbingers of even greater changes to come. When these Devonian Period fossils were first brought to the attention of the scientific community in 1915, it was suspected that they were the earliest evidence of reproduction by means of seed, a hypothesis later proven in the 1980s. The claw-like ends to the branches, called cupules, are receptacles that hold four seeds within, each one slightly smaller than a grain of rice. These first seeds opened the way for plants to exert greater control over their reproduction, leading to the development of highly varied accessory organs that eventually evolved into features familiar to us as flowers.

Seeds, technically called ovules before fertilization, are large spores that are retained on and nurtured by the parent plant. In living species, they are wrapped up in one or more protective layers of parental tissues, which were only partially formed in the earliest fossils. Pollination happens before the ovule is dispersed when pollen grains land on the surface and begin the process of fertilization, changing ovule to seed. In this way, seed plants reduced their reliance on films of surface water for their swimming sperm, a requirement in seedless cousins like mosses and ferns, during fertilization. By evolving seeds, a plant provisions its offspring in preparation for the crucial establishment of the next generation. For this purpose, seeds are often enriched in starch, making them attractive foods. Around the developing seed, plants began to evolve structures that facilitated pollination and dispersal. The function of the cupule in these early fossils is still very much a mystery; one attractive idea is that as wind blew across the open end it created a vortex that helped concentrate and funnel pollen towards the unfertilized ovules within. Much later on, seeds became associated with structures that attracted and rewarded insect pollinators, like showy petals and glands that produce sugary nectar. Tissues also associated with seeds can develop into fleshy structures that also attract and reward animal agents of dispersal. By such elaborate devices, plants have been able to recruit many varied animal allies to their cause.

# Keeping stems upright
## *Rhynia gwynne-vaughanii*

Rounded or many-sided cells of different dimensions make up this circular section through the stem of a plant, revealing the hidden nature of its internal tissues. We know that this plant was capable of growing on land, supporting its own body weight, and thriving in an environment where water is a limiting resource. The stem is less than 2 mm in diameter, yet within this small space we already see cells of varied shapes and functions. This is the most ancient plant for which we have such complete information, and it provides insights into how plants began to regulate their water usage and thereby to live a life on land.

The cells in the centre make up the vascular system (i.e. xylem, phloem) that specialized in transporting water and sap vertically though the stems. They are narrow in cross section, but when viewed from the side they prove to be very long. The cells around the outside of the stem are much shorter; they are the epidermis that formed a thin barrier between the interior tissues of the plant and the atmosphere. Those in between were most probably involved in photosynthesis, but they also played a role in stiffening the stem to keep it upright. The water essential to the life of the plant was drawn in from the soil through a rudimentary root system and then transported up through the stems by means of the narrow, pipe-like cells of the xylem. Perforating the epidermis, but not visible in this cross section, were numerous tiny pores called stomata that could be opened and closed to regulate the exchange of water and gasses with the atmosphere.

The remarkable preservation of the cells and tissues in this 407 million year old fossil plant results from a very particular set of circumstances. The plant once grew in a geothermal wetland near a hot spring that periodically flooded the landscape with water heavily charged with minerals. Once inundated, the plants drowned and, as the water cooled, chemical elements precipitated from solution to form the mineral opal within the plant cells and the neighbouring sediments. Thus, at this geological site in Scotland, a time capsule was forged that preserves in faithful detail the plants and other organisms that lived together at the dawn of life on land. This exceptional fossil shows that the physical processes driving water transport were similar in these early plants to those in modern species, including capillary action and the controlled evaporation of water from the surface of the stems mainly through the stomata. It is less certain, but also seems likely, that sugars made by photosynthesis were transported through the narrow phloem-like cells of the vascular system. The basic mechanisms of water and solute transport crucial to most modern species were among the earliest innovations of plant life on land.

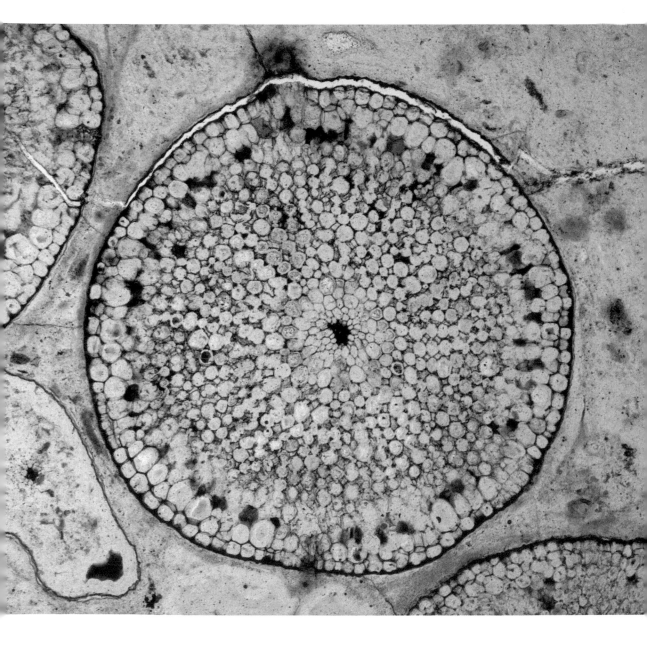

# Roots make soil
## *Stigmaria ficoides*

Hauled up from beneath a seam of coal 90 m (295 ft) below ground, this rock bears a short section of the massive rooting system of an extinct tree. The main shaft of the root has a speckled appearance; each circular black dot marks a small depression that is the point of departure of a long, narrow rootlet. These are most clearly observed to the right of the shaft, but they would have emerged all around its circumference, giving the root a bristly appearance. These roots were shallow and extended laterally for many metres from the upright trunk of the tree. They are common fossils everywhere that coal from the Carboniferous Period is mined.

Roots have been called the hidden half of plant evolution. Penetrating through the soil, and mostly concealed from view, they are vital in the lives of plants. The rooting systems of the first plants resembled those of modern liverworts and mosses. In these plants, stems are anchored to the ground by minute filamentous cells called rhizoids. Large plants require something more substantial. A key innovation was the evolution of specialized rooting branches that grew laterally or downwards, binding sediments to form deeper soils and penetrating cracks and crevices in rocks. The development of roots established both a stable anchoring platform and a network of branches that could scavenge water and nutrients from the soil, channelling them upwards towards the trunks and leaves above the ground. As plants grow there is a tendency for proportional relationships to be preserved between their above-ground and below-ground organ systems, thus larger trunks and canopies have larger rooting systems. Large roots were a prerequisite for the evolution of the trees that populated the ancient forests of the coal-forming swamps over 310 million years ago. The evolution of roots also had environmental consequences. Roots have important physical and chemical effects on soils and on the underlying rocks. Their presence enhances the natural weathering of calcium and magnesium silicates, which are abundant minerals in rocks at the surface of the Earth. These silicate minerals are weathered by rainwater, which contains carbon dioxide gas dissolved from air in the form of dilute carbonic acid. The weathering reaction creates soluble carbonates that eventually flow into rivers and oceans. Scaled up to a global level and measured over millions of years, the effect of root-enhanced weathering is gradually to reduce the concentration of carbon dioxide gas in the atmosphere and thereby reduce greenhouse effects causing climate cooling. Roots are one of several critical biological influences on the cycling of carbon and phosphorus through rocks and the biosphere, contributing to the habitability of our planet in the longer term.

# First trees
## *Eospermatopteris*

This bulbous base, with its distinctive elongate grooves, is the trunk of a tree from one of the world's most ancient fossil forests. A small grove of tree trunks was discovered in the Catskill Mountains, New York State in 1869 and subsequent excavations during the 1920s led to the unearthing of many more specimens. These discoveries were a serendipitous by-product of the growing demand for water in New York City, which led to the creation of the Schoharie Reservoir. Large-scale quarrying operations to construct the Gilboa Dam exposed the fossil-bearing deposits. Named *Eospermatopteris*, the tree trunks vary greatly in size and shape. Some are found standing in an upright position and have a bulbous base, which is typically broken off about a metre above ground. Other fallen trunks are longer and somewhat flattened. Because only the trunks were preserved and the internal tissues were mostly replaced by sediment, the true nature of these trees remained a mystery for many years, spawning various speculative reconstructions. A series of recent discoveries now provides answers, telling us about how these 380-million-year-old trees grew and shedding light on the accompanying flora and fauna of the first forests. The initial breakthrough came in 2007 with the discovery at another site in Schoharie County of a long, fallen trunk that bore a tufted crown of leaf-like foliage. This was the first firm evidence for the upper part of the tree. Viewed from a distance, the foliage superficially resembled the leaves of ferns, but closer inspection shows that instead of a flat, expanded blade, which is typical of modern species, the leaves were finely divided and filamentous. The presence of scars along the trunk showed that the foliage was ephemeral, being shed on a regular basis. Like ferns, the plant reproduced by means of minute spores, but unlike all living ferns these developed in clusters of tiny elliptical sacs at the ends of the leaf-like filaments. The trunks were anchored to the ground by numerous narrow roots that flared out from the base. It is estimated that these trees could reach a height of 8 m (26 ft), and with their pole-like trunks, topped by a tuft of foliage, they bore a passing resemblance to living tree ferns, cycads and palms.

Further evidence of the nature of the trees came to light in 2017, but this time from trunks discovered in the Xinjiang Uygur Autonomous Region of China. Unlike the specimens from New York State, these were petrified in silica, which preserved their internal tissues. The Chinese fossils gave the first real insights into the way that the trunks functioned. Instead of a dense, homogenous, woody interior, that would be typical of modern hardwood and softwood species, the amount of wood was very limited, confined to a mesh of narrow bundles near the circumference that braced and supported the weight of the tree. The bulk

of the trunk was composed of soft, non woody tissues and its girth increased by scattered cell division within the non-woody parts. Roots were also an integral part of the trunk. They began their development towards the periphery, growing outwards and then down the outer surface to form a covering, or mantle, of roots that was thickest at the base. This dense network of shallow roots anchored the trunk firmly to the ground. It is the mesh of narrow woody bundles and the well-developed root mantle that impart the distinctive grooved appearance to the surface of the fossilized trunks.

The *Eospermatopteris* trees were unique in the way that they grew; yet, they show parallels with modern species like palms, suggesting functional similarities. The internal construction of the tree trunks indicates that they had a high degree of flexibility; like palms, they would have been able to bend a long way without snapping. The crown of feathery leaves would have folded in a gale, like an inverted umbrella, reducing resistance to the wind, while the root bundle tethered the trunk firmly to the ground. These properties indicate that *Eospermatopteris* behaved in a similar manner to modern palms in the face of winds and storms. Over the course of their geological history, plants have found many varied ways to make trees, each of which differs in its biomechanical properties. Broadly similar solutions have also evolved repeatedly and at different times.

The fossil forest at Gilboa was home to several distinctive types of small tree, of which *Eospermatopteris* was the largest. During a brief period in 2010, repairs to the Gilboa Dam required removal of backfill from one of the early quarries, exposing 1,200 m² (12,900 ft²) of the original floor of the fossil forest. The sizes and positions of some 200 trees were recorded, giving a real sense of forest composition. The flora included scrambling herbaceous plants, mostly distant relatives of modern clubmosses. Woody lianas with narrow flexible stems grew up the trunks of the large trees. At other sites nearby, closer inspection revealed the legs and bodies of tiny arthropods protruding from beneath the plant remains. These proved to be fossilized fragments of their exoskeletons, which were originally made from the highly resilient polymer chitin. The fauna was dominated by arthropods, including centipedes and millipedes, pseudoscorpions, bristletails, mites, and the ancestors of spiders. Most of these were living in the leaf litter and the soil; some as predators and some feeding on decomposing plant detritus. Despite the presence of animals, there is little evidence that the living plants in these early forests suffered from herbivory, which was something that became important during later geological periods.

# Ancient forests
## *Lepidodendron*

Overlapping scales form a pattern with a natural twist on the surface of this broken branch of a fossilized tree. Large trunks bearing curious patterns of this sort were commonly found in coal-mining regions during the nineteenth century, where they were occasionally put on display, and advertised as enormous fossilized serpents. Although not animals at all, the trees from which these fossils came are as spectacular and alien as any mythical serpent or dragon. Called *Lepidodendron*, from the Greek meaning scale-tree, they grew upwards of 45 m (148 ft) in height bearing a high crown of simple branches. In life, each scale was the point of attachment of a long, narrow, grass-like leaf, imparting a bristly appearance to the surfaces of trunks and branches. These trees grew in coastal swamps over vast swathes of the equatorial regions of the Earth, giving rise to enormous deposits of coal that fired the Industrial Revolution and remain a significant energy resource today.

Between 325 and 280 million years ago, during the latter part of the Carboniferous and early Permian periods, most of the Earth's continents were aggregated into a single, nearly continuous landmass, in which the areas that today are modern Europe, North America and Asia were situated closer to the equator. Here, low-lying coastal land supported great forests that grew under tropical conditions. These forests are among the best understood of all ancient ecosystems due to intensive coal mining that has unearthed thousands of species of fossil. By far the largest plants were the scale-trees, which grew in flooded swamps. Their towering pillar-like trunks with bristly branches and long, narrow leaves would have cast little shade in the tropical sun. Far below the canopy of the scale-trees stood a denser understorey of smaller trees and shrubs dominated by plants with fern-like foliage. Some of these were true ferns, whereas others belonged to an extinct type called seed ferns. As the name implies, seed ferns reproduced by means of seeds; in contrast, the true ferns and scale-trees were spore producers like their modern descendants. Seed ferns are extinct relatives of the conifers, cycads and flowering plants (see p. 79). Much like wet tropical forests today, the small trees were festooned with epiphytes and climbers that were capable of rooting into their fibrous trunks or twining their way up by means of hooks and tendrils. In drier parts of the landscape ancient relatives of the horsetails prevailed. Today, and in the past, horsetails are easily recognizable by their jointed appearance and whorls of fine branches. Their underground rhizomes enable them to spread so that they form dense clumps. Below this lower canopy, on the shaded forest floor, were carpets of herbaceous ferns and clubmosses. Most of the Palaeozoic coal that we burn today originated in

these forests as carbon dioxide gas that was converted into plant organic polymers by photosynthesis. Under the flooded conditions of the swamps, as plants died or their parts were shed they formed a peat. Further subsidence and burial in the Earth's crust subjected the peat to heat and pressure, altering its physical properties and chemical composition. The plant remains were compressed, water and other volatile components were driven off, increasing the relative proportion of carbon and thereby the potential energy yield per unit weight. These formed the seams of coal that have been mined as a high-quality source of fuel for centuries, especially in eastern North America and Europe.

The coal forests were home to a varied fauna that included early relatives of living amphibians and reptiles, many of which would have been rather salamander-like in general appearance. They had slender bodies, blunt snouts, sharp teeth, short stout limbs projecting at right angles and a tail. They occupied shallow pools and leaf litter where they were opportunistic predators, subsisting on a diet of arthropods and fish. Many and varied land-dwelling arthropods also lived in the leaf litter, including such familiar forms as millipedes, centipedes, bristletails and silverfish. Primitive spiders, some of which lacked silk-producing spinnerets, ambushed or chased down their prey rather than catching it in webs. Scorpions already packed a sting in their tail like their modern relatives. Insects were the first animals to evolve flight, taking to the air without competition from birds or bats. These early fliers included species that are related to cockroaches, mayflies, dragonflies and damselflies. The Paleodictyoptera, on the other hand, were an entirely extinct order of medium to large insects. Their diagnostic trait was an additional pair of small wing-like appendages mounted in front of their true wings. They are thought to have fed on plants, ingesting pollen, spores, seeds and sap. Some of the arthropods greatly exceeded the maximum size of their living relatives. *Meganeura monyi*, for example, an extinct relative of present-day dragonflies, had a wingspan the size of a small seagull. Its body was over five times the length and twice the width of the largest living dragonflies. Millipedes in the genus *Arthropleura* could be more than 2 m (6½ ft) long, almost six times the size of their largest living relatives. It is thought that these gigantic forms could arise in the coal forests because concentrations of oxygen in the atmosphere rose to an unprecedented high, perhaps reaching 30%, greatly exceeding today's value of 21%. This was possible because of the burial of organic carbon in soils, which represents an excess of photosynthesis over respiration. The formation of coal deposits on such a vast scale had impacts that were felt worldwide.

# Ways to make a tree
## *Psaronius brasiliensis*

Polished agates can make beautiful display specimens, exhibiting some striking patterns and colours. This roughly circular piece, measuring about 20 cm (8 in) across, has a buff-coloured inner zone enclosing distinctive bands with scroll-like margins. The outer zone is made up of numerous small ovals in varied colours that radiate outwards. These clearly defined shapes and intriguing geometries were formed by the growth of a tree fern, and this is a section through the trunk. The inner zone is the stem and the distinctive bands with scroll-like margins of varying size represent different developmental stages of the vascular system that fed the leaves growing at the crown of the trunk. Surrounding this, in the outer zone, each small oval is a section through a root. The roots formed a dense covering, or mantle, that grew down from the top of the trunk enveloping and buttressing the inner stem. The different colours in the agate have no biological significance. They are caused by small quantities of minerals, mostly oxides of iron, that were deposited during the fossilization process.

This fossil from the Permian Period illustrates one of many mechanical solutions that plants have evolved to enable them to grow tall, which confers many advantages. Height helps in the search for sunlight by lifting the leaves above competitors, thereby reducing shading. Large plants can produce more offspring whilst providing an advantageous platform for their dispersal. Large size often goes hand in hand with longevity, meaning that the reproductive lifespan of individuals is extended. But being large has its costs too. There needs to be a significant and sustained investment in structural materials, and it can take time to reach sexual maturity. Also, all large plants start off small, so the benefits of large size are not immediately realized. The trunk is a key element. This needs to be strong enough to bear a crown of leaves or branches but flexible enough to withstand the sheering effects of wind. From the standpoint of efficient use of energy and resources, a good trunk should require little in the way of maintenance and it should be as cheap as possible to construct. Also, it must self-assemble and be fully functional throughout the development of the individual. Plants have found many different solutions to these problems, each of which brings its own advantages and constraints.

In most modern trees, trunks are woody cylinders; this is typical of conifers and many broad-leaved trees. Here the trunk is formed through the continual addition of wood to the interior and bark to the exterior, gradually expanding the girth. This neat solution results in a structure that is highly resistant to buckling and which easily grows

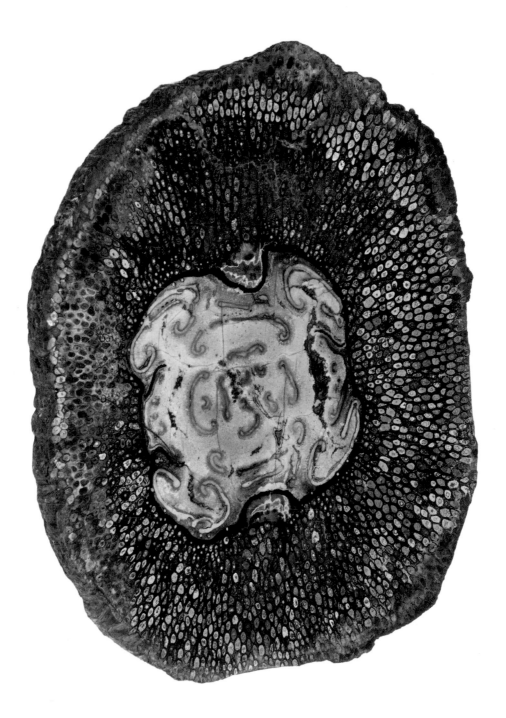

incrementally during the life of the individual. Plants built in this way can grow to great size, branch profusely, and develop a spreading crown. The tallest trees on Earth today are constructed in this manner, some coastal redwoods (*Sequoia sempervirens*) exceeding 100 m (328 ft) in height. Fossils like *Psaronius* and modern tree ferns adopted a completely different yet very elegant approach to constructing a trunk. These plants have no true wood. Their strength comes from the fibrous mantle of roots that envelops the central stem and the petioles leading to the leaves. Because the roots grow downwards from the apex, the root mantle acts like a buttress that is thickest at the base of the trunk. The trunks of tree ferns are strong, very lightweight, fibrous, and able to absorb and retain rainwater for crown growth. Because they have cross-linked strengthening elements, they are more sensitive to bending and to forces acting perpendicular to the long axis. This means that few tree ferns branch at all in the crown and they seldom exceed 20 m (65 ft) in height. They develop into small trees that are more suited to the understorey, where they have some protection from wind.

Plants have evolved many strategies in their quest to grow tall, and the most unusual and surprising are those of the strangler figs in the genus *Ficus*. The seeds of these figs are first carried into the forest canopy by bats or birds, where they germinate on the branches of tall trees. The young fig sprouts leaves that grow upward to catch the sunlight, while its roots grow downwards, some enveloping the trunk of its host, others descending towards the forest floor. Dirt and organic matter lodged in crevices in the host tree's bark provide a ready, though meagre, source of water and minerals during the fig's early life. Once the descending roots take hold in the soil, the growth of the fig quickens and its true intent becomes apparent. Its roots branch to create a mesh that envelops the trunk of its host in a deadly embrace. The fig now begins slowly to throttle the host tree by squeezing its trunk. Not all strangler figs kill, but some do. The roots continue to grow and, like those of the tree fern *Psaronius*, they form a buttress that eventually enables the fig to stand on its own feet. By the time the supporting tree dies, the strangler fig has become an independent tree, with its own crown of branches and leaves. Sometimes, the trunk of fused roots so completely envelops the body of its host that it leaves no outward sign of its sinister past. Tree-like growth forms have evolved over and over again in plants, as have reversal to the herbaceous habit. What suits one plant under one set of conditions is not optimal for another. Plant habit has shown enormous flexibility through the ages, responding to the ever changing ecology of the environments in which plants live.

# Conifers dominate
## *Agathoxylon*

This fossil comes from the Painted Desert in Arizona, USA, which is known for its vividly coloured landscapes in which bedrock in varied hues of red is exposed as laterally extensive layers on sparsely vegetated hillsides. In its southeastern quarter the surface of the desert is strewn with enormous logs of petrified wood from the Triassic Period. The logs are resilient, enduring on the surface as the sands and clays in which they were entombed are eroded away by wind and rain. Eventually, they tumble down slopes fracturing into smaller pieces, and gradually breaking up into innumerable tiny pebbles. These logs are the battered and worn trunks of trees that have lost their bark, limbs and roots. They were transported by river before fossilization, and during their journey their outer surfaces were scraped and worn. The giant limbless trunks became waterlogged, sank, and were slowly buried in sediment. Before the sediment was compacted petrification took place when silicates dissolved in pore water transformed into a type of quartz. In most cases this mineral completely replaced the original woody interior of the trunks. Mixed in with the typically pale quartz were small quantities of iron minerals and trace elements like cobalt, chromium and manganese that produced vibrant colours similar to those of the entombing desert sands.

The petrified tree trunks of Arizona are massive, averaging 24–30 m (79–98 ft) in length, with some exceeding 40 m (131 ft). It has been estimated that the tallest trees would have reached 59 m (194 ft) in height with a trunk 3 m (10 ft) across at the base. The big trees were conifers. It is possible to deduce this from the cellular structure of the wood, which is composed of small cells of uniform size, where this has been preserved. Most trunks lie on their sides but, occasionally, the base of one is found upright and rooted in a fossil soil. From this we know that the roots were robust, and that they were adapted for growth under wetland conditions in deep, water-soaked soils. Preservation of the bark is exceptionally rare because it falls away after death and is relatively easily abraded. Where found intact it is thin, which is consistent with the humid, frost-free tropical climate that is thought to have prevailed at the time. The growth habit was probably somewhat similar to living species like the Douglas fir. The wood is attributed to the genus *Agathoxylon*, also formerly known as *Araucarioxylon*. These names hint at an affinity with the distinctive southern hemisphere family Araucariaceae, which today includes the monkey puzzle tree, the wollemi pine and the kauri. However, this is misleading because fossil woods of this type are typical of a much broader range of living and extinct species. To narrow

OPPOSITE PAGE: A tree trunk preserved in the Petrified Forest National Park, Arizona, USA from the Triassic Period, 225 million years ago.

down further the nature of these trees and how they are related to known families of conifers needs additional information about their leaves, cones and seeds. In the Arizona forest these parts of the plants have never been found in connection to the trunks, so these spectacular fossils have not yet given up all of their secrets.

The Mesozoic Era (251–66 million years ago) was the age of the conifers, and the abundance of fossil wood in the geological record testifies to their prominence. Conifers were the largest trees, and they dominated many ecosystems until the rise of flowering plants towards the close of the Mesozoic Era. In today's woodlands and forests species of conifers are vastly outnumbered by broadleaved, flowering trees. Yet, despite their low species diversity, conifers are still ecologically important in many regions of the world. Larch, spruce, fir and pine dominate the taiga or boreal forest of the northern hemisphere, which is the world's largest terrestrial biome. In regions that have mild winters and heavy rainfall temperate coniferous forest is common and conifers can attain truly massive sizes. Conifers also grow under conditions of seasonal aridity in the Mediterranean climate of North Africa and southern Spain. Today they remain prominent in some of the tropical and subtropical montane forests of the Americas, Indomalaya and China.

The modern families of conifers can trace their origins to the latter part of the Mesozoic Era, but living species represent only a part of their true diversity. It is quite possible that the large fossil trees of the Arizona forest belonged to an entirely extinct lineage. Evidence from the DNA of living species now tells us that much of the modern species diversity of conifers originated more recently during the Neogene Period (23–2.6 million years ago). This pattern is most marked in the northern hemisphere, where it seems to be related to a general long-term trend of climate cooling. About 40 million years ago warm tropical or subtropical climates were widespread at latitudes that are temperate today. Since then, the northern hemisphere has become cooler and drier, favouring the spread of frost-tolerant lineages. More recently, cycles of glaciation led to repeated contractions and expansions of species ranges, fragmenting populations and encouraging speciation. Different effects were felt in the southern hemisphere due to the disposition of continental landmasses. Here relatively warm and wet refuges endured, often moderated by oceanic climates, leading to the persistence of older conifer lineages. The modern distributions of conifers and the pattern of their species diversity were greatly shaped by climate change on a global scale influenced by the distribution of the world's continents and oceans.

# Forests in the Arctic
## Axel Heiberg Island forest

The fossil forests that once existed in the high Canadian Arctic are among the most northerly ever recorded, standing in stark contrast to the treeless landscape of today. These relics of earlier warmer times tell a story of climate change on a global scale, and they provide intriguing insights into high-latitude forests that have no modern analogues. They are so remote that they were only discovered in 1985, when a helicopter pilot, flying for the Geological Survey of Canada over the remote Axel Heiberg Island, noticed an unusual concentration of logs and stumps protruding from an otherwise barren ridge. On closer inspection these proved to be the remains of fossil trees preserved in their original growth positions and looking almost as fresh as the day that they died. Located at about 80°N, this forest lies well within the Arctic Circle.

The Axel Heiberg Island forest, estimated to be about 45 million years old, dates to the middle part of the Eocene Epoch. The wood is preserved as lignite, meaning that it is in the earliest stage of coal formation yet still retains much of its original integrity. A saw not a hammer is needed to collect these fossils. The trees are rooted in a peaty soil from which one can pick up leaves and cones. The structure of the wood and the associated fossils enables us to establish the identity of the trees. Most are extinct species, but closely related to the living dawn redwood (*Metasequoia glyptostroboides*) and the Chinese swamp cypress (*Glyptostrobus pensilis*). Growth rings in their wood indicate that the trees grew under a warm and moist climate. There was an understorey of broadleaved flowering trees and shrubs, and ferns on the forest floor. This was a wetland community similar in some respects to the Florida Everglades, with alligators, crocodiles, turtles and even an extinct hippopotamus-like mammal called *Coryphodon*.

The Axel Heiberg Island forest provides one of many strands of evidence that Earth was warmer 45 million years ago. Despite the influence of plate tectonics, this particular part of the Arctic is thought to have remained at constant latitude since the Eocene Epoch. This means that the forest community was exposed to extreme seasonal variation in day-length unlike that experienced by any forests today. The plants endured three months of continuous summer light and three months of continuous winter darkness. The trees and shrubs were deciduous, becoming dormant to survive the winter. Plant growth in the Arctic today is limited more by the cold, dry winds and nutrient-poor soils than by winter darkness. These fossils affirm that the annual variation in sunlight is not a limiting factor to the development of forests even at this high latitude.

# Nutrients for plants
## fossilized fungi

These two flask-shaped structures represent complementary views of a fossilized fungus that was found in 407 million year old soil. Smaller than a grain of pollen it was imaged using a microscope that emits laser light, causing the fungus to fluoresce. To build up the image, different planes of focus were captured and then further processed to render surfaces in false colour. The flask is conical with an elongate neck, and the remains of hyphae can be seen projecting from three points on its surface. The outer surface has been made transparent in the image at the bottom to reveal that it contains numerous even tinier spores. These and other details tell us that this is a zoosporic fungus. Today these fungi are predominantly decomposers. However, some are pathogens, the most familiar being the causative agent of chytridiomycosis, which is a widespread infectious disease of amphibians.

Together with bacteria and other micro-organisms, fungi are essential and ubiquitous elements of the soil microbial community. Even though zoosporic fungi are aquatic, they also thrive in the moisture found around soil particles. We now know that such fungi played a vital role in early freshwater ecosystems and wet soils. Their enzyme systems are capable of breaking down resilient organic materials like cellulose, chitin and keratin. They remobilize essential elements, including carbon, nitrogen and phosphorus, recycling them to sustain the growth of plants and other organisms.

At the same ancient geological site in Scotland one finds other fungi emerging from pores in the surfaces of plant stems or preserved within their tissues. Some of these were probably pathogens. Occasionally, one can observe reaction responses by the plant, which attempts to erect barriers to the fungus or to seal it off into compartments through necrosis of selected tissues, the thickening of cell walls, and the formation of encasement layers. Other fungi appear to be able to enter plants without provoking an immune response. Their hyphae easily slide between cells weaving their way into the interior. Once they are a few cells deep, they change tack to form distinctive branched or coiled structures within cells. These are mycorrhizal fungi. A plant may allow such a fungus to infect it so that a partnership can be established. The plant benefits because the fungus is more efficient at extracting essential nutrients from the soil through its extensive network of hyphal filaments. In return, the plant rewards the fungus with sugars it manufactures by photosynthesis. These intimate partnerships are ancient and may well have been essential to the initial establishment of plant life on land.

# Carbon cycle
## rotten log

It has been said that there is more life in a dead tree than in a living one. Dead wood is vitally important to the health of a forest, providing habitats for an enormous range of organisms including small vertebrates, invertebrates and many varied mosses, lichens and fungi. Under natural conditions, it can contribute up to a quarter of above ground forest biomass. This log was part of a large branch that fell in a forest 145 million years ago. As the wood dried it began to shrink, losing its outer layer of bark and becoming infested by fungi and wood-boring insects. As the rot progressed, the wood around the outside began to lose its integrity and to break into cubical pieces. Before the rot was complete, this particular forest suffered a catastrophic flood that buried the fallen logs in layers of silt and clay. Silicates dissolved in the floodwater eventually precipitated inside the wood as a type of quartz, turning wood to stone and preserving this partially rotted log for posterity.

In temperate forests biological decomposition of a fallen tree typically takes 50–100 years. The resilience of wood is due to the nature of its cells, which are made of cellulose fibres bound together with other carbohydrates and a complex aromatic polymer called lignin. These are large molecules constructed from repeated subunits that are difficult to break down chemically. The capacity to digest wood is very rare among animals. Beetles and termites that feed on wood are mainly reliant on enzymes produced by bacteria and protozoa in their guts. Certain species of bacteria have the capacity to rot wood on their own, but decomposition is comparatively slow. The principal and most aggressive decomposers of wood today are Agaricomycete fungi, the group of fungi that includes mushrooms, puffballs, bracket fungi, boletes and chanterelles. While not all species attack wood, and some even form mutually beneficial symbiotic associations with plants, many species possess enzymes that are capable of degrading cellulose and lignin to cause rots.

One of the vital effects of these communities of decomposers is to slowly recycle the carbon locked up in dead wood back into the atmosphere as carbon dioxide gas, where it is once again taken in by living plants and converted to sugars by photosynthesis, resuming the cycle of life. Recent research on the genomes of fungi is revealing how their rot-inducing enzyme systems evolved, and these seem to post-date the first evolution of wood. So, whereas this ancient log has suffered attack by fungi and wood-boring insects, wood in the most ancient forests would have been subject to different and as yet unknown communities of decomposers.

# Plants as food
## *Trigonocarpus parkinsoni*

A prominent theme in the ecology of life on land is the interactions between plants and the arthropods that feed on them. Today these are mostly insects, but also included in their number are many mites, myriapods and some crustaceans. The survival of an estimated one million or more insect species depends directly on plants as a source of food. Plant-eating arthropods employ varied and sophisticated feeding strategies, and the plants respond through the production of myriad chemicals that exert toxic or repellent effects on their animal attackers or which reduce their value as a source of food. This intense struggle that has shaped the evolution of both groups has a lengthy geological record.

Arthropods were the first animals to exploit plants as food, a relationship that began over 400 million years ago. Initially, this involved feeding on easily ingested or digested materials, progressing later to the more resilient tissue systems. Fossil evidence encompasses nibbling marks on the edges of leaves, galls, stem borings, gut contents and faecal pellets. Plant spores swallowed whole were an early source of nourishment for small arthropods, and they are often abundant in fossilized faecal pellets. The resilient spore walls passed through the gut, but their cell contents were more readily digested. Some tiny arthropods evolved piercing mouthparts that tapped straight into the plant's vascular system to feed off sugars, much like modern aphids. Another ready source of food was the remains of dead plants. This detritus had the advantage of already being partly decomposed by fungi and bacteria, making it easier to digest.

The predation of seeds came later. As seeds evolved their storage reserves of starch, oils and proteins, intended for the developing plant embryo, became a tempting high-quality source of nourishment. Each of the two fossil seeds shown here is about the size of a plum stone. Each has distinctive ridges, but the one at the top also has a circular pore that bears a striking resemblance to the borings made by insect larvae in modern seeds. The last items to appear on the menu were leaves, wood and roots. Their tissues have a high cellulose content and contain aromatic hydrocarbons making them challenging to eat. Furthermore, these tissues are often marinated in chemicals that have deterrent or toxic effects. Insects developed sophisticated chewing mouthparts and recruited many kinds of microbial symbionts to enable the digestion of these plant tissues. By the time that the coal-forming forests were flourishing, about 310 million years ago, arthropods had already evolved many feeding strategies to exploit the new opportunities afforded by plants. This is a struggle that never ceases as insect herbivores continue to battle the ever-evolving resistances of plants.

# Insect pollination
## *Pegoscapus*

Amber from the Dominican Republic preserves an insect that became trapped in resin exuded from a tree 25 million years ago. Examination, using confocal laser scanning microscopy, revealed granular clusters with a golden hue just visible near the mid region of the body cavity (bottom image). These are pockets of fig pollen, similar to those carried by living fig wasps. Plants in the genus *Ficus* (figs) have a distinctive inflorescence comprising a hollow ball, the interior of which is lined by numerous tiny flowers. Only female fig wasps can enter by crawling through a narrow pore to lay their eggs inside, pollinating the fig flowers at the same time. Fig wasps and fig trees are mutually dependent, with each of the approximately 750 living species of fig typically pollinated by one or a handful of species of tiny wasp. Pollination by animals happens in 87% of all living species of flowering plants. The pollinators are overwhelmingly insects, but few have developed relationships as tightly integrated and as interdependent as the fig and its wasp.

Insect pollination has had a tremendous impact on the evolution of flowers and flowering plants, but its origins are much older. It probably has its roots in the consumption of pollen and spores by insects, which pre-dates pollination. There is a growing body of evidence that insect pollination first originated in the gymnosperms. Modern conifers and *Ginkgo* are wind pollinated, but in many cycads pollination is affected solely by beetles, and pollination by flies is inferred for some species of *Ephedra* and by moths in *Gnetum*. During the Mesozoic Era, curious features associated with reproduction in some extinct gymnosperms suggest insect pollination. These include unusually large pollen grains, glandular hairs, accessory tissues and the orientation and position of ovules within cone-like or flower-like reproductive units. Many fossil insects have mouthparts that are consistent with pollination. Beetles were probably among the oldest pollinators since they have biting and chewing jaws capable of consuming pollen. Other insects possessed straw-like protrusions adapted to pierce pollen grains or to siphon, sponge or lap up fluids exuded from within or around ovules. Added to this is the direct association of gymnosperm pollen as a dusting on very early fossil insects. It is therefore likely that many Mesozoic Era gymnosperms were pollinated by insects, including thrips, lacewings, flies and beetles, which are also important pollinators today. Many of these lineages of gymnosperms subsequently became extinct during the early evolution of the flowering plants. Some of their pollinators probably shifted hosts to flowering plants, but the rise of flowers was also associated with many new pollinators including the butterflies and moths and the bees and wasps.

# The herbivores
## coprolites

Scattered across the surface of this rock are more than 40 flattened, beige-coloured discs with irregular outlines. They are somewhat variable in size, averaging about a centimetre in diameter. These are coprolites or fossilized faecal pellets that were made by an animal about the size of a large rabbit or perhaps a sheep during the Jurassic Period. Adhering to many of the coprolites are black organic patches. When these were macerated in a strong oxidizing agent they were found to be composed of fragments of plant cuticles. The cuticles bore distinctive cellular patterns enabling them to be recognized as belonging to the leaves of a small cycad-like plant that was common in the associated flora. The animal that produced the coprolites was thus a herbivore that had recently fed almost exclusively on this shrub. The fossilization of the gut contents of herbivores or their faecal pellets is very rare, but when this happens it provides direct evidence of animal diets. One cannot know for sure what animal made these faecal pellets, but in rocks of this age the most common herbivores of this size were dinosaurs.

The earliest vertebrates to emerge onto land were amphibious carnivores that fed on arthropods and fish. As life on land became more firmly established some animals began to incorporate plants as part of their regular diet, or as a seasonal resource, evolving into omnivores. Herbivory evolved subsequently and on many occasions, typically from ancestors that were omnivorous. In living vertebrate animals, herbivores occupy a wide dietary spectrum ranging from species with high-fibre diets of leaves, stems and bark to those that feed selectively on fruits and seeds that are richer in starches and sugars. The transitions to herbivory entailed the development of a complex set of adaptations to a diet rich in cellulose. The most prominent of these involve modifications to the skull, teeth and digestive system, and these are most marked in the high-fibre herbivores. Some dinosaurs evolved cranial and dental adaptations that rival those of any living mammal. These included batteries of interlocking teeth, high tooth replacement rates, horny beaks and complex jaw mechanisms to shear and chew their food. The evolution of herbivory is also frequently associated with a substantial increase in body size. Animals that browse or graze need an enlarged digestive system within which to nurture and contain the symbiotic bacteria that are essential to the breakdown of cellulose in the plant fodder. Plants have left their mark on animal evolution in many ways, but the numerous switches to herbivory has led to some of the most rapid and spectacular diversifications in animal life, including the evolution of the largest animals to live on land.

# Giant clubmosses
## *Stigmaria*

Among the most distinctive and impressive of fossil tree trunks are the pillars of stone that are frequently encountered during the mining of coal. The image shows a tree stump with part of its massive intact rooting system. It is a natural sandstone cast measuring a little over 1.7 m (5½ ft) in height and weighing in excess of one tonne. A close relative of *Lepidodendron* (see p. 38), the plants that formed these trunks were among the largest trees of the coal swamp forests, with some attaining heights of 45 m (148 ft).

This specimen was discovered in an open cast mine in North Wales, UK, but similar trunks can be found in deep mines hundreds of metres below ground where they pose hazards. They are often encountered in the shale above a coal seam during tunnelling operations. Observed from underneath they appear in the roof as a thin ring of coal, which marks the outer surface of the trunk, enclosing a circle of sand or clay. Because of their appearance, they were commonly known as cauldron bottoms. When undermined they are secured in place with a bolt, because there is an alarming tendency for the whole trunk to slide out of the roof as one massive column, crushing miners and equipment in the tunnel beneath. *Stigmaria* belongs to the lycopods or clubmosses, which during the early evolution of life on land were among the most varied and abundant species of plants.

Today the clubmosses and their kin are small, obscure plants that make up a tiny fraction of the modern flora. *Lycopodium* can be found in light woodland or on open ground. It forms extensive networks of prostrate creeping stems with small needle-like leaves that give rise to erect branches bearing pale yellow cones. *Selaginella* is common in tropical forests and an invasive weed of greenhouses. Many species have leaves and stems flattened into frond-like branches. The closest modern relatives of the ancient Carboniferous Period trees are the quillworts in the genus *Isoetes*. Their long, quill-like leaves are the most conspicuous of several striking similarities that have persisted over the eons to furnish clear evidence of kinship. However, unlike their giant forebears, these are very short plants with a tufted growth form. Today many species are aquatic or semi-aquatic, growing in freshwater lakes or thriving in waterlogged conditions. Among the quillworts and their fossil relatives there has been a great reduction in size through time. Although size and growth form have changed radically in these archaic plants, they have tenaciously held their ground in wetlands and marginal aquatic habitats for over 300 million years.

# Before flowers, ferns
## *Cladophlebis australis*

This fragment of fern frond from the Walloon Coal Measures of southeastern Queensland, Australia, exhibits a striking and rare form of preservation. The delicate veins and outlines of the leaflets are traced in a pale opaline silica, which stands in contrast to the brown organic remains of the leaf blade and the maroon colour of the encasing iron-stained rock. The leaves are rather similar to those of living ferns in the Osmundaceae, or the royal fern family, to which the plant is probably related. Before the evolution of the flowering plants, a little over 100 million years ago, ferns like these were among the most abundant and varied elements of the herbaceous ground cover. They dominated open landscapes, rather like grasses and herbs do in modern prairies. Many families of ferns trace their origins to the Mesozoic Era or earlier, yet others are of more recent origin. There is evidence that most of the more than 10,000 species of living ferns evolved alongside the flowering plants, perhaps adapting to grow in the more varied habitats that they created, including heavily shaded forest floors and the trunks and branches of trees that now host numerous epiphytic species.

It should not surprise us that ferns have as complex and as varied an evolutionary history as any other ancient group of plants. However, for some brief intervals of time their abundance eclipsed all other species. One well-studied example is the first few centimetres of rock of the Palaeocene Epoch, about 66 million years ago. Here, the floras are frequently dominated by one or a few species of plants, usually ferns. So notable is this phenomenon that it is referred to as the 'fern spike', because of the rapid and short-lived abundance of fern spores observed in these rocks. Fern-rich fossil assemblages of this sort are interpreted as evidence of environmental recovery following a catastrophic event. Today ferns are early colonizers of disturbed environments, particularly in the tropics. They readily invade volcanic landscapes and burned ground following forest fires. Their ability to colonize quickly and in large numbers is due in part to their buried rhizomes that readily survive fire as well as their numerous tiny spores that are capable of being carried great distances on the wind. This brief flourishing of ferns in the early Palaeocene followed the worldwide environmental catastrophe that led to the extinction of the dinosaurs at the end of the Cretaceous Period. Possible causes include asteroid impact and volcanism on a massive scale related to the development of the Deccan Traps – a vast accumulation of igneous rocks in central India. Events of this scale had a devastating effect on plant and animal life. The first green shoots to re-sprout were those of the ferns, heralding the opening of the Cenozoic Era.

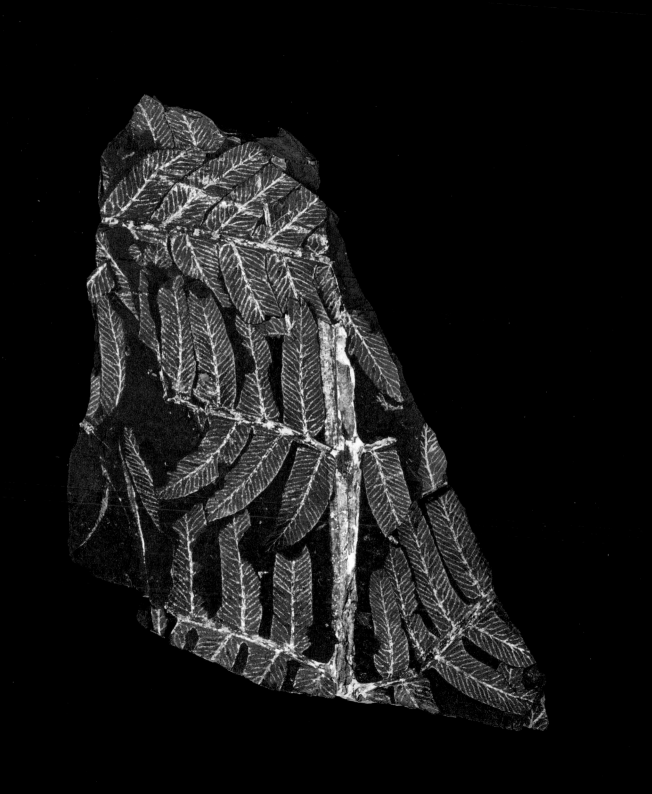

# Thriving horsetails
## *Asterophyllites* and *Palaeostachya*

Delicate jointed stems bearing simple lateral branches in whorls are common fossil in coal-bearing rocks of the Carboniferous Period. In this specimen the lateral branches take on two forms, one is narrow and flexuous with simple bristles, and the other is thicker, stiffer in appearance and has pointed ends. Known as horsetails, there are only 15 living species all classified in the genus *Equisetum*. Modern plants are weedy, herbaceous perennials with upright aerial stems that arise from a very extensive underground rhizome. They have a very distinctive and conservative form that is easily recognizable. Each main stem bears sheaths of small scale-like leaves; in many species, whorls of narrow branches also develop at the nodes. Reproduction is by means of spores that develop in cones. Plants range in size from an 8 m (26 ft) tall tropical species to a diminutive temperate one that rarely exceeds 5 cm (2 in) in height. Today *Equisetum* is an isolated and impoverished group, but horsetails have a lengthy and varied fossil record.

Horsetails are commonly regarded as living fossils, a term Charles Darwin originally coined in 1859 in reference to the platypus and the South American lungfish. A living fossil lacks an agreed and succinct definition, but it is a term applied to species or groups of species that diverged from their nearest living relatives tens or even hundreds of millions of years ago. It is also applied to survivors of once large groups that have few living species, and to distinctive organisms that apparently have undergone little change in their form and ecology over millennia. On the face of it, horsetails seem to fit all of these criteria.

Some of the earliest fossils that are clearly related to horsetails come from Bear Island in Spitsbergen, Norway, in rocks of the Devonian Period that are over 380 million years old. Their affinity with the living forms is immediately apparent in the jointed appearance of their stems and the whorls of branches developing at the nodes. They differ from modern *Equisetum* sufficiently to be classified in the separate genus *Pseudobornia*, and they were much larger, growing to more than 15 m (49 ft) in height. Horsetails were especially prominent in the peat-forming swamps of the Carboniferous Period with growth forms that varied from herbaceous climbers to large trees. The upright trunks of *Calamites* were woody, measured up to 20 m (65½ ft) in height, and developed from an equally massive underground rhizome. The fossilized parts of the plant are given different names. *Asterophyllites* is a common name for the foliage and *Palaeostachya* is a type of cone, which is shorted and stouter than the foliage. The fossils illustrated could therefore be parts of a much larger plant. Like modern *Equisetum*, the ancient *Calamites* favoured wet or waterlogged soils, thriving

on loosely consolidated substrates such as sand bars at the margins of lakes and streams. The earliest fossils attributable with confidence to *Equisetum* come from 136-million-year-old sediments of the Cretaceous Period from British Columbia. These fossils bear a strong general resemblance to the smaller living species. Today *Equisetum* is widely distributed, mostly in temperate climates of the northern hemisphere. Recently they have benefited from the development of roads and railways, which have created significant new habitats in their associated drainage ditches. One anomaly in their modern distribution is the absence of native species from New Zealand and Australia. These regions host a rich and varied fossil record of horsetails, and in New Zealand fossils have been documented in rocks as recent as the Miocene Epoch. The cause of their demise remains puzzling. One idea is that it is related to substantial environmental changes associated with the rifting of New Zealand and Australia from Antarctica and their subsequent northward migration from high to lower latitudes. This entailed significant shifts in climate and soil moisture regimes. Competition with the native flowering plants during the Neogene Period probably also played a role.

Horsetails thus fit the general criteria of living fossil, but today use of this term is contentious because it appears to subvert post-Darwinian evolutionary thinking by implying that organisms remain unchanged for millions of years. Others argue that it leads to a progressivist worldview, in which some organisms are viewed as somehow higher or more advanced than others. In horsetails, aspects of their form and ecology are undoubtedly similar to their Mesozoic Era relatives, but this should not be taken to imply that they are unchanged in other respects. The 15 living species are the sole surviving link to a once ecologically prominent and varied group of plants. One recent study of their family tree based on DNA evidence showed that all of the living species can trace their ancestry back to a surprisingly recent common ancestor that lived during the early part of the Cenozoic Era. Furthermore, most of the living species evolved even more recently during the Miocene Epoch. It can be concluded that much of the modern species diversity in this ancient group is of comparatively recent origin, even though many aspects of their appearance and structure seems to have remained unchanged for much longer.

Scale of one foot.

Untreated, the nut-like seeds of *Cycas media* are poisonous. Indigenous peoples of Australia developed a method of removing the toxins rendering the seeds edible. From *Botanical Drawings from Australia* (1801) by Ferdinand L Bauer (1760–1826).

extinction during the evolution of the cycads. Furthermore, it points to the Neogene Period as the proper geological context for understanding the origins of their modern species. One plausible influence on species evolution during this time is the general trend towards climate cooling and increased seasonality in the tropical and subtropical regions favoured by most cycads. Despite their recent burst of evolution, it is estimated that two-thirds of living species are at risk of extinction, with some already extinct in the wild.

Large cycads resemble palms, but they are not closely related, since palms are flowering plants, whereas cycads are related to conifers. Likewise, cycads also resemble extinct plants in the Bennettitales (see p. 88), whose fossils are so similar to cycads that the leaves and trunks of the two are often impossible to tell apart. Adding to the confusion, when fossils cannot be attributed with confidence to either group they are informally termed 'cycadophyte', meaning cycad-like. This does not mean that they are cycads. Detailed studies of the reproductive organs and other anatomical features of cycads and Bennettitales show that they are not closely related plants. As with palms, their similarities in habit are an example of convergent evolution. The illustrated fossil has been attributed to both cycads and Bennettitales, and in this case there is simply not enough information preserved to decide which attribution is correct.

Cycads evolved before the dinosaurs and have long outlived most of them, yet the two co-existed for much of the 185 million years of the Mesozoic Era. Did herbivorous dinosaurs feed on cycad foliage and perhaps even on their seeds? There is little direct fossil evidence from stomach contents or faeces. Turning to evidence from the modern world, all living species of cycads contain virulent toxins that can cause severe debilitating symptoms and even death in mammals, although some species can eat them without apparent ill effects, showing a preference for young, tender leaves. Cycad seeds are often brightly coloured. In most species, the kernel is rich in starch but highly toxic, whereas the fleshy seed coat is edible. Animals that feed on cycad seeds typically strip away the edible coat and discard the toxic kernel or they swallow the seeds whole, digesting the seed coat and allowing the kernel to pass through untouched. Cycads therefore discourage browsing animals but recruit and reward others that disperse their seeds. Hornbills and parrots are known to feed on cycad seeds, and birds are now recognized as a specialized subgroup of theropod dinosaurs. So, even though there is little direct evidence that *Iguanodon* or any non-avian dinosaur fed on cycads, it is possible that the foliage and the seeds were once part of the diet of those remarkable extinct animals.

# Cones of conifers
## *Araucaria mirabilis*

Some fossil plants are so dazzling that they take on mineral or gemstone-like qualities. This conifer cone is preserved as chalcedony, a type of microcrystalline quartz, and it was created by tectonic processes that led to the formation of our modern southern hemisphere continents. About the size of a lemon, it has been cut through with a diamond-bladed saw and polished to reveal large seeds of a similar shape to pine nuts arranged around a central axis. The fossil formed during the Jurassic Period when South America experienced volcanism on a truly massive scale. This was the beginning of the break up of Gondwana. Enormous volumes of volcanic rock were deposited. Ash was ejected from vents in such great quantities that it killed and preserved whole forests. The results can be seen today in the Cerro Cuadrado Petrified Forest of Patagonia. This is an arid region with little vegetation, yet the barren landscape is littered with massive petrified tree trunks, twigs and cones. During the Jurassic Period it was a coniferous forest that was dominated by trees of the ancient family Araucariaceae.

Araucariaceae typically are very tall evergreen trees with an impressive and distinctive habit. There are about 37 living species, mostly in the southern hemisphere. The monkey puzzle tree, *Araucaria araucana*, is a widely planted ornamental, and it is by far the hardiest species in the family, with a tolerance of cool temperate climates and abundant rainfall. Araucarias were important in Mesozoic forests. Like most conifers they are wind-pollinated, and they have two types of cone borne on separate plants. The seed cones of araucarias – and of conifers in general – are more varied than their pollen cones, and this is related to their different functions. Pollen cones evolved simply to produce and release large quantities of pollen into the wind, so their shape and structure are quite conservative. Seed cones play a greater variety of roles, balancing the protection of developing seeds with the need to disperse mature ones, and facilitating pollen capture. One begins to see the onset of more varied forms of conifer seed cone during the Jurassic and Cretaceous Periods, and this seems to be related to increasing interactions with dinosaurs, birds and mammals. The evolution of large, tightly packed cones with many seeds like those of araucarias and pines afforded a measure of protection against predation, whereas the simplified cones that form a fleshy, brightly coloured berry-like fruit, as seen in podocarps and yews, were adapted for dispersal by birds. Although quite varied, the diversity of conifer seed cones does not match that of flowers, perhaps because unlike flowers insect pollinators have never played a major role in shaping their evolution.

# Transitional fossil
## *Archaeopteris hibernica*

Transitional fossils are extinct species that bridge gaps in the hierarchy of life. To us they can appear bizarre because they possess surprising combinations of features. The term 'missing link' is often applied, but palaeontologists tend to avoid such usage because it gives the impression that life is a linear hierarchy, a great chain of being, when in fact evolution is better depicted as a tree with many diverging branches. There are numerous examples of transitional fossils in animal evolution, and plants have their own too. At first sight, botanical examples may not be readily apparent because fossil plants typically are found as isolated organs such as wood, fruits or leaves. The strange combination of features only come to light when different pieces of plant can be convincingly shown to come from the same species. This magnificent fossil called *Archaeopteris* measures over half a metre in width. At first it was thought to be a fern, but its true nature remained unrecognized for nearly 100 years.

In 1871 the Canadian geologist Sir John W. Dawson illustrated fossil fern-like fronds from the Devonian Period that he attributed to *Archaeopteris*. Doubts about their relationship with ferns were raised in 1939 by the American palaeobotanist Chester A. Arnold who discovered that *Archaeopteris* reproduced by means of spores of two different sizes, which is very rare in living ferns. Earlier, in 1911, an apparently unrelated discovery by the Russian palaeobotanist Mikhail D. Zalessky was later to prove significant. Zalessky described a new type of petrified wood from the Donets Basin, Ukraine. He called this wood *Callixylon*, and although he did not find any foliage or reproductive structures attached to the trunks, Zalessky noted the fossils' similarity to modern conifers. Subsequently, trunks of *Callixylon* have been found at many sites throughout the northern hemisphere, where they are often associated with the fossil fronds of *Archaeopteris*. In 1960, almost 100 years after Dawson first described its leaves, the American palaeobotanist Charles B. Beck made the unexpected discovery that the foliage (*Archaeopteris*) and the wood (Callixylon) were actually parts of the same plant.

Plants that combine fern-like leaves, and reproduction by spores, with a trunk composed of very conifer-like wood do not exist today. This transitional fossil reveals the existence of hitherto unknown plants that were intermediate between ferns and the seed-producing plants that eventually gave rise to conifers, cycads, ginkgo and flowers. Its unique combination of features raises many questions because there are no close modern analogues. We know that these fascinating plants grew as trees and shrubs in early forests, but many aspects of their biology are still shrouded in mystery.

# Long extinct seed ferns
## *Physostoma elegans*

Some forms of fossilization preserve small patches of the environment with such fidelity that they are analogous to time capsules. So-called 'coal balls' are in this category. These are concretions made mostly of carbonates of calcium and magnesium that formed within the peaty soils of coal swamp forests 315 million years ago. In this cut section of coal ball, fragments of plant tissues form the backdrop to a distinctive, pale, oval body measuring about 2.5 mm. It is bounded by thin layers of tissue of varying thickness, density and texture, and finger-like projections emerge from its outer surface. This is a tiny seed that is clothed in secretory tissues and robust tubular hairs. The cells in the seed coat are clearly discernible, and just barely visible within there appear to be numerous minute bubbles. These are the spore-producing bodies of fungi that invaded the seed to feed on its nutritious interior. Seeds of this type developed in the seed ferns, which are a long extinct group of prominent understorey plants of ancient forests.

Fern-like foliage is extremely abundant in the rocks of the Carboniferous Period, but by the turn of the twentieth century palaeobotanists suspected that many of the leaves did not actually belong to ferns but rather to a novel extinct type of seed plant. One clue came from features of the foliage that were puzzling. Unlike ferns, the leaves were produced on woody stems and the spore capsules that one would naturally expect to find on the underside of their leaves had never been observed. To establish their true nature reproductive organs were needed, but finding these was difficult. In 1902 Francis W. Oliver made a crucial breakthrough when he noticed that part of a seed-bearing organ, preserved in a coal ball, bore distinctive glands that were very similar to glands on a leafy stem. Although a physical connection could not be established, the similarity of the glands was good evidence linking seed to foliage. Seeking further evidence, Oliver engaged the help of an energetic young research assistant, Marie C. Stopes, who made additional crucial observations of the glands. Oliver joined forces with Dukinfield H. Scott to publish the discovery, confirming the existence of a major new group of plants. The seed ferns are not ferns. They resemble ferns in the general shapes of their leaves, which is one of many examples of convergent evolution among plants. Inspired by her role in this discovery, Marie Stopes went on to complete a doctoral thesis and to become a palaeobotanist. Later her career took a different turn, and she is now better known as an author, social reformer and vocal campaigner for women's rights, opening Britain's first birth control clinic in London in 1921.

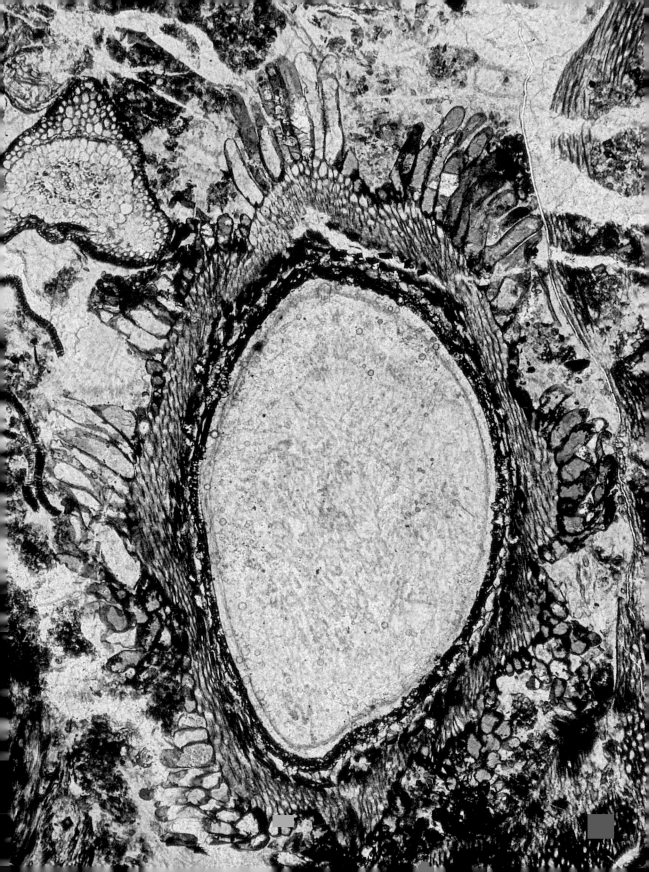

# Back from the brink
## *Ginkgo cranei*

This elegant leaf is one of many that were blown off a shrub or tree into a lake over 50 million years ago, eventually becoming infused by waters rich in minerals. Each of its delicate veins, picked out in pale, opaline silica, traces a forking path that radiates out of the long, narrow petiole into the fan-shaped blade. Microscopic examination reveals that the leaf's waxy cuticle preserves an exact imprint of the pattern of cells in its epidermis. Also found preserved in silica in the same sediments are small seeds, some of which retain their original spherical form, revealing a thick, fleshy coat enclosing an apricot-like stone. These features are hallmarks of the genus *Ginkgo*, commonly known as the maidenhair tree. The single living species, *Ginkgo biloba*, originates in China, but it is now widely cultivated across the world as an ornamental, renowned for its tolerance of urban settings under temperate climates and for the beauty of its foliage. In parts of eastern Asia the nuts are a delicacy, and they are used in traditional medicine. The tree's distinctive leaves and great longevity have taken on symbolic meaning. It is revered in China, Korea and Japan, where magnificent old trees stand over ancient temples and shrines. Because of its long history of cultivation, the existence of truly wild populations in China remains an open question.

*Ginkgo* has a geological history reaching back at least 200 million years. The fan-shaped leaves are very distinctive, but they were more varied in the past. The earliest species that are clearly attributable to the ginkgo group had leaves that were divided into many lobes, and the seeds were smaller and more numerous. *Ginkgo* rose to prominence becoming most varied and cosmopolitan during the Jurassic and the early part of the Cretaceous, where it has been found on every continent, including Antarctica. Around 100 million years ago its success came under challenge by the newly evolving flowering plants that began to invade *Ginkgo*'s preferred streamside habitats. Striking and dynamic patterns of distribution emerged during the Cenozoic Era that seem to be shaped by the ecology of *Ginkgo* and its response to climate change. *Ginkgo* is first noted as missing from tropical floras whereas it remained abundant in temperate floras, signalling a preference for cooler climates with marked seasonality. This pattern is most notable during the Eocene Epoch, when the Earth's climate was much warmer and the tropics extended to higher latitudes. During the Oligocene and Miocene epochs the tropics contracted and cooler, drier climates spread across mid latitudes. *Ginkgo* followed, moving out of its high latitude refuge to spread south again in Europe, North America and Asia. In the

southern hemisphere it was a different story. Presumably outpaced by the extent and nature of environmental change, *Ginkgo* disappeared completely, perhaps as early as 24 million years ago. As cooler climatic conditions spread over the northern hemisphere, woodland gave way to prairie in North America and to steppe in Asia, further restricting the distribution of *Ginkgo*. Yet, fossils tell us that *Ginkgo* grew naturally in the wild 15 million years ago in Iceland and just 5 million years ago in Bulgaria and Greece. The ice ages of the Pleistocene took a heavy toll, and by their end *Ginkgo* had vanished from all but a handful of protected valleys in eastern and south central China. By the time modern humans arrived, about 50,000 years ago, *Ginkgo* was a relict. Peter R. Crane, in his book devoted to *Ginkgo*, makes the point that its recent geological history serves to remind us of how the world was a very different place not so long ago. Viewed from a geological perspective, climate change can have a profound influence on the distribution and variety of plants over relatively short time intervals.

Climate change during the Cenozoic Era played a huge role in the demise of *Ginkgo*, but it is not the whole story. *Ginkgo*'s ecological preferences might have rendered it especially vulnerable. The fossils point to a long association with rivers and wetlands, which are environments that are prone to change. Rivers shift their courses and water levels fluctuate. Coastal wetlands are particularly sensitive to changes in sea level. Other gymnosperm trees with similar ecological preferences, including the Japanese umbrella-pine (*Sciadopitys verticillata*), the Chinese swamp cypress (*Glyptostrobus pensilis*) and the dawn redwood (*Metasequoia glyptostroboides*), have also experienced great reduction in their distributions in the recent geological past. Seed dispersal might also have been a weakness. *Ginkgo* has separate male and female trees. The fruits are large, 2–3 cm (¾–1 in) in diameter, with a distinctive odour of rancid butter, but little is known about the natural dispersal agents. One intriguing theory proposes that the original animal dispersers became extinct, thereby reducing the capacity of *Ginkgo* to respond to environmental change. If this is correct, *Ginkgo* became stranded, tied to its own ground, without a means of escape in the face of climate change. This is a difficult theory to test, but long-distance dispersal of large heavy seeds is better developed in plants that encourage feeding by birds or bats. *Ginkgo* was brought back from the brink of extinction through its appeal and utility to us. Its survival is assured, so long as it continues to serve *Homo sapiens*, its newly acquired friend and modern agent of dispersal.

# The lonesome pine
## *Agathis jurassica*

Talbragar is one of the most significant freshwater fossil deposits of Jurassic age in Australia. Its pale silicified plants preserved in an ochre-coloured shale also make it one of the most recognizable. The plants were fossilized, alongside fish and insects, in the quiet bottom-waters of a lake 150 million years ago. It can be difficult to identify plants with certainty based only on knowledge of their leaves. These leaves were once likened to the kauri (*Agathis*), a genus of about 22 living species of tropical conifers in the Araucariaceae family. Today's experts are sceptical and would rather attribute the foliage to the fossil genus *Podozamites*, meaning that its relations with conifers are unclear. Conifers were prominent elements of Mesozoic Era floras, yet much about their past diversity still remains unknown. Even today, we occasionally make surprising discoveries. The dawn redwood (*Metasequoia glyptostroboides*) was thought to have gone extinct during the Pliocene Epoch, until a single living species was discovered in Hubei Province, China, and brought to the attention of botanists in 1945. In 1994, a new tree in the family Araucariaceae was discovered only 150 km (93 miles) from Sydney, Australia's most populated city. Like the Talbragar fossils, its foliage caught the eye, but it was not immediately recognizable, even to experts. A return visit to collect leaves, bark and cones proved decisive. The new tree was named the wollemi pine, *Wollemia nobilis*. It was not just a new species of Araucariaceae, it combined features of the two living genera *Agathis* and *Araucaria*, suggesting that it belonged to an ancient lineage that had diverged from other living species a long time ago.

The Wollemi National Park is a protected area of wilderness in the northern Blue Mountains and Lower Hunter regions of New South Wales. It is a rugged, impenetrable landscape dominated by deep valleys, canyons, cliffs and waterfalls, formed by the weathering of ancient sandstones and basalts. The wollemi pine was discovered by David Nobel, an officer with the National Parks and Wildlife Service of New South Wales, and two friends. While abseiling into a narrow canyon they noticed a 35 m (115 ft) tall tree with peculiar bark and an archaic appearance towering above the more familiar trees of coachwood (*Ceratopetalum*), sassafras (*Atherosperma*), lilly pilly (*Syzygium*) and possumwood (*Quintinia*). Their discovery turned out to be the biggest of its kind since *Metasequoia*. Wollemi pines grow in natural coppices, with multiple trunks developing in clumps. They have distinctive, dark brown, knobbly bark. The leafy branches dip gracefully and appear feather-like from a distance. Seed-bearing cones are large and spherical; pollen cones are smaller and slender, developing on the lower branches. Incredibly, only three

copses of this remarkable tree survive, and genetic studies show that individuals are nearly identical, indicating that they are natural clones. Because of their rarity the exact location of the wollemi pines is a closely guarded secret. They are at risk from unscrupulous collecting, and they are vulnerable to pathogens inadvertently brought into their inaccessible home by visitors. Hunkered down in their deep ravines, the wollemi pines cling on, protected from the seasonal aridity of the fire-prone eucalyptus forests that today dominate this wilderness.

The geological history of the wollemi pine and its close relatives is best told by its pollen. In 1965 a new type of fossil pollen grain was recognized based on its distinctive granular coat and named *Dilwynites*. At the time the plant that produced this pollen was unknown. With the discovery of *Wollemia* the mystery was solved. *Wollemia* pollen closely resembles *Dilwynites*, providing a handy means of tracking the rise and fall of the wollemi pines in the geological record. *Dilwynites* is widely found dispersed in sedimentary rocks, first appearing in Australia 90 million years ago and later in New Zealand, South America and Antarctica. From about 34 million years ago, its fortunes reversed. *Dilwynites* shows a steady decline before it finally disappeared from Australia about 2 million years ago. The pollen record therefore points to a long geological history for the wollemi pine, but one that does not extend back as far as the Jurassic Period. Any resemblance between the Talbragar fossil foliage and *Wollemia* is superficial. These leafy fossils are neither *Wollemia* nor its sister genus *Agathis*, but they could be a more distant extinct cousin.

The wollemi pines were once abundant and widespread trees that grew in moist, high-latitude southern hemisphere forests. Their decline and near extinction in Australia has been attributed to progressive drying of the climate, accompanied by an increase in wildfires, which is linked to the development of ice sheets in Antarctica and to powerful tectonic forces that caused Australia to drift northwards following its break from Antarctica. Inevitably, this picture is now not quite as clear as once thought. Recently two living species of *Agathis* were also shown to have pollen like that of *Wollemia*, so the distribution of *Dilwynites* through time might be telling a bigger story. Like the Talbragar leaves, fossil pollen is often not sufficiently diagnostic to attribute to a single living species or genus. Today, *Wollemia nobilis* is critically endangered. It is protected by efforts to conserve it in its natural habitat. It also benefits from conservation programmes that cultivate threatened species. *Wollemia nobilis* is now the most frequently grown endangered conifer in arboreta worldwide.

# Extinctions
## *Monanthesia saxbyana*

This pleasingly patterned rock shows the surface of the trunk of a small shrub. It is formed of silica, so to enhance its features it was etched with hydrofluoric acid. The acid treatment revealed a regular pattern of dark, deeply etched triangles. Each one marks the attachment of a leaf stalk, now long gone. Just above most triangles are much smaller faceted shapes that are seemingly clustered in whorls or spirals, some more distinct than others. These are the organs of reproduction, which in this plant are embedded in the trunk. They are flower-like cones consisting of a whorl of pollen-bearing scales enveloping numerous ovules or seeds, but only the tips are visible. This type of reproductive structure is unique to an extinct group of small, Mesozoic Era trees and shrubs called the Bennettitales. In their general habit, they resembled living cycads, but similarities are superficial. The two groups are not close relatives. Bennettitales are one of the many casualties of plant evolution. They became extinct towards the end of the Cretaceous Period. Their trunks are also among the earliest fossil plants known to humankind. A specimen was discovered in a 4,000-year-old Etruscan tomb, perhaps placed there as a token by a loved one attracted by the symmetry of its patterned surface.

Extinction appears to be the ultimate fate of all species. It is hard to be sure, but recent estimates judge the natural baseline extinction rate to be about one species for every ten million species per year. On five occasions over the past 540 million years of Earth history the rate of extinction soared to notable highs. Numerous species and even major groups of organisms disappeared over relatively short intervals of geological time, changing the character of life on Earth. These are called mass extinctions, and it is no accident that four of these mark transitions between major geological periods. Perhaps the most famous is the Cretaceous-Palaeogene extinction that saw the demise of the non-avian dinosaurs, about 66 million years ago. The most lethal was the Permian-Triassic extinction, which wiped out many animal groups completely about 252 million years ago, permanently changing the ecology of marine ecosystems. Mass extinctions are caused by colossal environmental change acting at a global scale and, in geological terms, over relatively short intervals of time. Their causes are debated, and they may have several. The Permian-Triassic extinction seems to be related to one of the largest known volcanic events of the past 540 million years. Over a geologically short interval of about 2 million years a massive volume of lava was deposited over an area that today covers 7 million km² (3 million miles²) of Siberia. The Cretaceous-Palaeogene extinction was likely caused by the impact of an

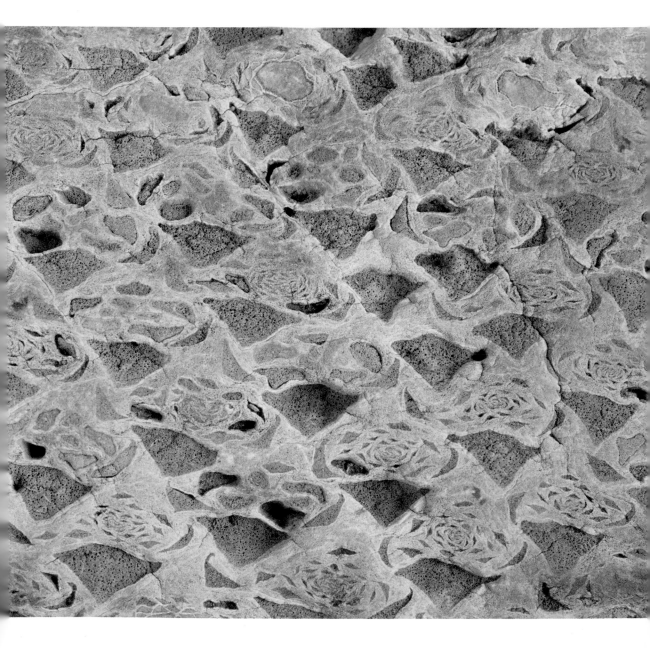

asteroid that created the Chicxulub crater on the Yucatan Peninsula in Mexico. Geologists first recognized mass extinctions by their effects on shelly marine animals, which are very varied and common fossils. Plants were affected too, but in other ways, reflecting fundamental differences in their ecology and life-cycles.

We might anticipate that the responses of plant communities to major environmental disturbance would differ to those of animals. Many plants are more resilient. They can regrow after damage, their seeds or spores can persist in the soil, and many species have a great capacity for dispersal. Also, plants are primary producers, so changes to plant communities would exacerbate the effects of environmental change on the animals that are reliant upon them. Studies of the effects of mass extinction on plants show that the impact comes through more clearly at a local or regional scale rather than at a global one. Plant species suffered, but unlike animals few higher taxonomic groups such as families were eliminated. Plant communities typically experience ecological upheaval, but such upheavals can happen at other times, too. During the latter part of the Carboniferous Period there is a major change in the nature of floras that seems to be driven by climate change but which does not appear to affect animal life significantly. Extinction can also be selective. Large and ecologically dominant species suffer more than smaller weedy ones. Insect-pollinated species seem to be more vulnerable than wind-pollinated ones. Unlike animals, extinctions of whole branches of the plant tree of life are much rarer. This is probably due to the greater ecological diversity seen among species within the families and genera of plants. Recovery of plant communities from the effects of mass extinctions is slow, taking millions of years. However, in contrast to what we see in animal communities, mass extinctions did not play a primary role in shaping the direction of plant evolution.

The Bennettitales and many other gymnosperms were already in decline towards the end of the Mesozoic Era, disappearing from certain parts of the world before the Cretaceous-Paleogene mass extinction. It is thought that they were displaced by the new and rapidly evolving flowering plants. There is some evidence now that Bennettitales together with the extinct seed ferns managed to endure into the Palaeogene Epoch in high-latitude pockets in the southern hemisphere, a further testimony to the resilience of plants to mass extinction. Studying the effects of extinctions in the past helps us better understand the likely scale and impact of human-driven environmental change in today's world.

# Latitude and diversity
## *Pagiophyllum peregrinum*

Plants and their leaves are in close equilibrium with the surroundings in which they grow. This twig has robust, diamond-shaped leaves that closely overlap, giving it a scaly texture. Each leaf also has a thick, water-repellent covering that was originally made of lipid polymers impregnated with waxes. Closer examination with a microscope would reveal that the minute pores, through which the plant exchanges gases and water with the atmosphere, are sunken in channels and partly obscured by finger-like papillae. These features help to reduce water loss under dry conditions. Plants evolve to adapt to the environments in which they live, leaving tell-tale indicators in fossils. When these are considered together with other fossil and geological evidence, they can provide remarkable insights into climates of the past and their effect on the distribution of vegetation on a global scale.

In today's world, species richness of vegetation generally increases from the poles towards the equator, peaking in the humid tropics. This pattern is so striking that it has been called the latitudinal diversity gradient, and it seems to hold for most land-dwelling organisms. Its causes are widely discussed but still not fully understood. However, what is typical today does not seem to hold for the past. One approach to investigating the geological distribution of plant diversity is to compare and contrast the published accounts of fossil floras from around the world. Add to this information on the distribution of coals, which form under conditions of flooding sustained by high rainfall, and the distribution of evaporites, which are minerals found in sedimentary rocks that formed under arid environments, and one can build up a picture of plant diversity in relation to climate and latitude. Studies of the Jurassic Period show that for plants the most productive and diverse regions on Earth were located at mid-latitudes, where the development of forests was extensive. Conifers dominated with an understorey of broadleaved plants including ferns, cycads and horsetails. Extinct plant groups like seed ferns and Bennettitales flourished. In striking contrast to today, polar regions were richly vegetated by large-leafed deciduous conifers and *Ginkgo*. Low latitude regions were much less varied, only patchily forested and more arid. Like *Pagiophyllum*, the plants that grew here were mostly small-leafed types with adaptations to life under water stress. Today's latitudinal diversity gradient probably reflects the waxing and waning of glacial cycles, which have shaped patterns of rainfall and temperature over the past 2.5 million years. Under warmer climates that prevailed for most of the past 500 million years, productivity and diversity were greater at mid-latitudes, and this seems to hold for land-dwelling animals too. Where plants go, animals invariably follow.

# Fruits and climates
## fossil fruits

These tiny fruits and seeds have a metallic lustre because they are formed of the iron sulphide mineral pyrite, which is also known as fool's gold. Their source is the London Clay, a geological formation that underlies a large part of southeast England, including London itself. They once grew on an ancient forested shoreline from where they drifted out into a shallow sea, eventually becoming waterlogged and sinking into the mud on the seabed. There, under anoxic conditions, sulphate-reducing bacteria went to work, eating away at the tissues and infiltrating cells and cavities with reactive fluids that produced the mineral. Fossil fruits of the London Clay have been collected and studied for more than 300 years, and over 350 species are named. In 1933 Eleanor Reid and Marjorie Chandler published a monograph, illustrating hundreds of species. Their careful descriptions of the fossils enabled comparisons with living plants. This showed that although all of the species are extinct, it is possible to classify many of them into modern genera and families. Overwhelmingly, the closest living relatives of these fossils grow today in the tropical forests of Southeast Asia.

During the early part of the Eocene Epoch (56–48 million years ago) several lines of evidence indicate that Earth's climate was warmer than it had been at any other time during the Cenozoic Era. Fossil plants tell us that southern England was subtropical rainforest with a mean annual temperature some 10°C (50°F) above today's. This is significant, even though the London Basin was 10 degrees further south at the time. The coast was fringed by mangrove swamps and the hinterland was a dense forest of trees, shrubs and lianas. Crocodiles and turtles inhabited its rivers and shorelines. The high latitudes reached by warmth-loving vegetation in the north enabled a great mixing of climate-sensitive plants among the Americas, Europe and Asia via land bridges across the North Atlantic and the Bering Strait. This flora, which spread across the northern hemisphere, consisted of an unusual mixture of tropical and temperate species, and its composition was strikingly similar throughout. The modern floras of Eurasia and North America largely derive from this source. Each region went on to develop its own distinctive character, influenced by subsequent geographic and climatic events. The widening of the North Atlantic Ocean eventually sundered ties between Europe and North America. The closing of the Turgai Strait, a seaway that separated Europe from Asia, facilitated exchange of species between these regions. The prolonged trend of climate cooling that followed the Early Eocene reduced the flow of species across the Bering Strait. This resulted in the contraction of the ranges of tropical species to their present distributions and the expansion of today's more familiar temperate flora.

# Freezing and the fall
## *Acer trilobatum*

Each year during the autumn the trees and shrubs of northern temperate forests put on a spectacular display of leaf colour before shedding their foliage and entering the season of winter dormancy. The character of these forests changes as the leaves that once made the canopy now carpet the floor and naked boughs shake against bitter winds. This maple leaf is one of many that were blown into the still waters of a shallow lake that formed in a volcanic crater about 13 million years ago. The various hues of brown might reflect variations in the intensity of original pigmentation that gave rise to autumn colours. On some leaves from this lake one can observe galls made by insects and mottled patches caused by fungal infections that are common on maple leaves during the late summer and autumn. The deciduous habit is a characteristic feature of northern hardwood forests where it is an adaptation to freezing temperatures. It also evolved in other settings that are subject to seasonal aridity, notably dry tropical forests. Deciduousness is one of many examples of convergent evolution in plants.

From a geological perspective, the deciduous habit of many flowering trees and shrubs evolved quite recently, probably in the latter part of the Miocene Epoch, when the modern distribution of cool temperate forests became established. The long-term global trend in climate cooling that began over 45 million years ago eventually led to annual periods of freezing temperatures in the mid-latitude forests in which this maple grew. Freezing can induce embolisms in wood, air bubbles that block hydraulic pathways, and are more likely to develop in woods composed of cells with large conduits, typical of many flowering plants. This type of wood is more efficient at transporting water to large, broadleaved canopies under warm conditions of plentiful year-round rainfall. The deciduous strategy works because it shuts down hydraulic function during cold periods, but there is a seasonal price to pay of shedding leaves and lost resources. A second strategy is to sacrifice hydraulic efficiency by building safer water-transport networks consisting of small-diameter conduits. Trees and shrubs that have taken this route are able to remain evergreen but at the cost of reduced productivity under ideal growth conditions during summer months. A third way was to change growth form and become herbaceous at the cost of losing leaves and stems aboveground and overwintering as seeds or underground storage organs. Today, the proportions of herbaceous species, deciduous species and evergreens with small water-conducting conduits increase towards the poles. The evolution of the deciduous habit probably happened gradually and independently in many hardwood species as a response to climate cooling over millennia.

# Flowers flee the Antarctic
## *Nothofagus beardmorensis*

Despite Antarctica being the most inhospitable continent on Earth, with its freezing climate and 4 km (2½ mile) thick ice-cap, some of the most common fossils preserved in its rocks are plants. These fossil leaves were found in the Transantarctic Mountains, approximately 500 km (311 miles) from the South Pole. They fell in a southern autumn, accumulating as mats in a glacial lake. Alongside the leaves are gnarled twigs. Their wood is light brown and soft enough to cut with a knife. Both leaves and twigs are from an extinct species of southern beech. Annual growth rings in the wood are extremely narrow, indicating a brief growing season of about 6–12 weeks that began following early summer snow melt and ended with hard autumnal frosts. Despite the twigs rarely exceeding one centimetre in diameter, ring counts establish that some plants were over 60 years old at time of death. They grew as dwarf trees in low tufts, hugging the ground to shelter from freezing winds, like the modern arctic willow. This tundra vegetation is the last known vestige of the original flora of Antarctica.

About 85 million years ago Antarctica was positioned roughly where it is today, yet many parts of the continent experienced subtropical climates. Summers were humid with high rainfall and temperatures averaging 20°C (68°F). There were probably no extended periods of freezing, even through the long, dark, polar winters. Flowering shrubs and herbs flourished alongside towering conifers and ginkgos, with many species related to families found today in the floras of South America, New Zealand and the islands of the Southern Ocean. Climate cooling led to the onset of glaciation about 45 million years ago. Warmth-loving plants gradually gave way to species that could tolerate freezing winters. Diversity fell, yet plants clung on tenaciously even after ice-sheets covered much of the land. Forests were eventually replaced by species-poor tundra communities dominated by mosses but including dwarf southern beech and conifers, grasses and sedges, and a handful of herbs. As the ice-sheets expanded, and the climate of the interior became increasingly cold and dry, this vegetation became extinct, replaced by a polar desert. Only two species of flowering plant are found on the continent today, and both are new colonizers that arrived within the past 10,000 years.

The age of these fossil leaves is difficult to determine and far from certain. Some authorities suspect that they could be as old as 17 million years, whereas others think that they are probably less than 3.8 million years old. The question of when the extinction of the flora of Antarctica happened hangs in the balance. It depends, in part, on determining more precisely the age of the rocks in which these fossils formed.

# Greening of the Sahara
## *Ficus*

This fossil formed when leaves fell into an artesian spring of water rich in dissolved lime that was deposited as tufa, coating objects with a white crust. Tufa deposits are found on a massive scale in the Western Desert of Egypt, which today is a hyper-arid region of the Sahara without natural surface water. It is a barren place of rocks and dunes, where the intense heat is relieved only by the relentless rasping of the sand-laden wind. It is uninhabited, save for a chain of oases, of which Kharga is the most southerly. Here, the aquifers that originally fed the tufa-forming springs are the only source of water for agriculture and daily life, but these waters now lie deep underground from where they are pumped to the surface. The flora and fauna within the tufa deposits testify to a time when surface water was abundant. The aquifer was brimming and spilled out into waterfalls, marshes, small ponds and streams. Rainwater accumulated in natural depressions forming extensive freshwater lakes, and the landscape was richly vegetated. These lime-encrusted leaves belong to species of figs that today grow widely across sub-Saharan Africa as shrubs and small trees.

A primary influence on the vegetation of North Africa is the extent and intensity of the West African Monsoon. Today this is confined to sub-Saharan Africa, but in the past this monsoon belt has shifted farther north, bringing increased rainfall that turns the Sahara green. This happened most recently between about 11,000 and 6,000 years ago. At this time, much of the Sahara became savannah, a mixed woodland and grassland ecosystem, supporting human populations throughout. The fossil leaves became trapped in tufa over 115,000 years ago capturing an earlier and even wetter green Sahara. Here, too, there is abundant evidence of occupation by our ancestors. Many such greening events are thought to have taken place over the past 8 million years. During its extended arid periods, the Sahara presents a formidable barrier, separating the fauna and flora of sub-Saharan Africa from Eurasia. But these periods of greening enabled interchange and, most significantly, opened routes for the migration of early modern humans out of Africa to spread across the world. The principal driver of these shifts in the monsoon belt is the gradual change in the tilt of the Earth's axis and the shape of its orbit, which affect the total amount of sunlight reaching different parts of the globe. The impact of this effect can be amplified or dampened by many other factors. One of these is the presence of vegetation itself. Plants reduce the amount of solar radiation reflected from the land surface, store water in their canopies and enhance soil moisture, contributing a positive feedback on rainfall. By their presence alone, plants can transform their habitats.

# Drifting continents
## *Glossopteris*

These spear-shaped leaves provide early clues to the existence of an ancient supercontinent called Gondwana. They are very common fossils in rocks of the Permian Period from India, Africa, South America, Australia and Antarctica. The leaves are large, and they are easily recognized by their distinctive shape and the presence of a strong midrib with a meshwork of finer veins. *Glossopteris* was a seed-bearing shrub or tree related to the gymnosperms. Occupying Gondwana's mid to high latitudes, many species were deciduous, which is one reason that the leaves are found in such abundance. These were predominantly swamp-dwelling plants, and their fossilized remains are the primary constituents of the major economic coal resources in the Permian rocks of India and the southern hemisphere.

The concept of Gondwana was conceived in the late nineteenth century based on striking similarities in fossil flora among the southern hemisphere continents. At the time the prevailing theory of the Earth saw continents as fixed in their relative positions, so kindred floras were explained by dispersal across ancient land bridges that had subsequently foundered. In 1912 this idea was challenged by the German meteorologist Alfred Wegener. According to Wegener, continents are not fixed, but over eons drift infinitesimally slowly, coalescing and subsequently fragmenting. Despite abundant circumstantial evidence in its favour – including the shapes of the southern hemisphere continents, geological continuities that crossed ocean basins, and similarities in fossil flora and fauna – the theory of continental drift lacked a convincing mechanism to drive movement through the Earth's crust. For this reason, it excited great controversy. Continental drift only became widely accepted in the 1960s, following further discoveries about the structure and processes governing the Earth, now embodied in the theory of plate tectonics. The rifting of Gondwana began about 180 million years ago in the Early Jurassic with the continents beginning to take their modern shapes and positions during the Cretaceous Period. This influential idea has been of paramount importance to understanding the modern distributions of plants on different continents. Explanations for these distributions can be complex, involving the sequence in which the elements of Gondwana broke up and drifted apart, dispersal of species over long distances and changes in Earth's climate on a grand scale. The role of Antarctica was also central, which is not apparent from the polar desert that we see today. Under milder climates in the geological past Antarctica was both a source of species and a land route through which they could disperse, uniting continents that are today separated by ocean basins.

# The leaf barometer
## *Ginkgo huttonii* stomata

Stomata are minute pores that are scattered across the surface of leaves enabling the exchange of gases and water vapour between plant and atmosphere. They function as valves, regulated by a pair of sausage-shaped cells, giving the appearance of tiny mouths. This fossil shows the pattern of epidermal cells on a fragment of leaf over 170 million years old. One can make out numerous stomata and other features, including small hairs. The shapes of cells are imprinted on the cuticle, which is a protective, impermeable film of lipid polymers and waxes that covers the surface of leaves. The cuticle is highly resistant to decay and is often preserved in fossils when all other tissues have gone. Epidermal features are a guide to the identity of the plant, but they can also give clues about the environment in which it grew. In some species, the proportion of stomata on the leaves varies with the concentration of carbon dioxide gas in the atmosphere. Plants use carbon dioxide gas in photosynthesis, but in carbon dioxide-rich atmospheres they require fewer stomata to meet their needs. Thus, the density of stomata can act like a barometer tracking the rise and fall of this gas in air through time.

This property of stomata was first noted in living species where it was tested in a natural experiment. Levels of carbon dioxide gas in Earth's atmosphere have increased by nearly 40% since before the Industrial Revolution. This effect can be measured directly in air bubbles trapped in cores extracted from ice-sheets in Greenland and the Antarctic. In 1987 the plant ecologist Ian Woodward examined the stomata on the leaves of seven species of tree and one shrub that had been collected from sites in England at intervals over the past 200 years and stored as dried specimens in an herbarium. In line with predictions, he observed that the proportions of stomata in the leaves of each species fell as levels of carbon dioxide gas in the atmosphere gradually rose from pre-industrial levels. Stomata are now used as a means of inferring changes in this important gas over geological intervals that extend far back in time beyond the ice-core record. Recent research shows that not all plants respond in the same way or to the same degree, but *Ginkgo* seems to be one that does. A common fossil, its leaves have been used to investigate the evolution of the atmosphere over hundreds of millions of years. Understanding the relationship between the concentration of carbon dioxide gas in the atmosphere and global climate is a matter of pressing concern today. This property of plants can help us to measure changes over long timescales and to investigate how the atmosphere has responded to major events in Earth history. Such knowledge will assist scientists in predicting how current rises in carbon dioxide may affect our climate.

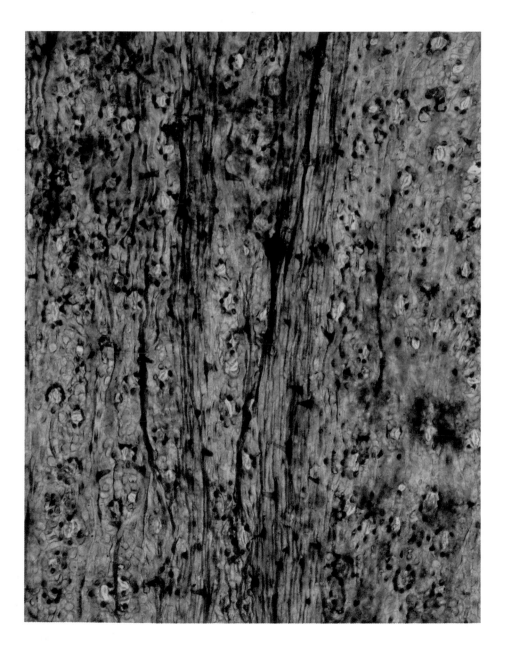

# Aquatic fern
## *Azolla*

Plants that evolve from land habitats into aquatic ones often undergo dramatic changes. This slab of rock captures part of a fern mat that was fossilized in lake sediments over 50 million years ago. Unlike typical ferns, this is an aquatic species that has a tiny body with scaly floating leaves and long, hair-like roots dangling freely below. Individuals could be plucked out of water on the tip of a finger, but collectively they formed massive bright green floating mats. Its living relatives are in the genus *Azolla*, the mosquito ferns, so called because their capacity for rapid growth in bodies of stagnant water reduces the ability of mosquitoes to lay eggs and hinders the development of their larvae. This fern also has a trick that makes it useful in agriculture. It forms a symbiotic association with a nitrogen-fixing bacterium that enables it to grow in freshwater where nitrates are scarce. These properties are put to work in Southeast Asia where *Azolla* is used as a green fertilizer in paddy fields. A useful ally today, this tiny fern had an enormous environmental impact at the time these fossils formed.

During the Middle Eocene Earth's climate was much warmer and there were no polar ice-caps. Because of the configuration of the continents at that time, the Arctic Sea was almost completely isolated from the world's major oceans causing unusual conditions to develop. Geological cores drilled through sediments deposited in this sea reveal extraordinarily high quantities of *Azolla* spores. Because living *Azolla* can only tolerate very slight amounts of salinity, this implies that freshwater conditions prevailed. It is thought that rainfall combined with water flowing into this sea from river systems formed a thin layer of freshwater that floated on stable briny subsurface layers. Under these unique conditions truly massive blooms of *Azolla* were able to expand across the surface of the Arctic Sea on a seasonal basis. Extraordinary phenomena like this can have an enormous environmental impact. The estimated area of the Arctic Sea during the Eocene was 4 million km$^2$ (1½ million miles$^2$), about the size of the European Union. Conditions favourable to the growth of massive blooms of *Azolla* are thought to have persisted for at least 800,000 years. As plants died, they sank to the stagnant sea floor where they were incorporated into sediments. In this way, carbon dioxide gas was drawn out of Earth's Eocene atmosphere and locked away in sea-floor sediments where it remains to this day. Although the extent of these blooms is debated, the removal of such large quantities of this gas from the atmosphere is thought to have made a major contribution to climate cooling on a global scale. Shortly after the *Azolla* spores disappear from the geological cores we begin to see the first evidence of glaciation on the other side of the world in the Antarctic.

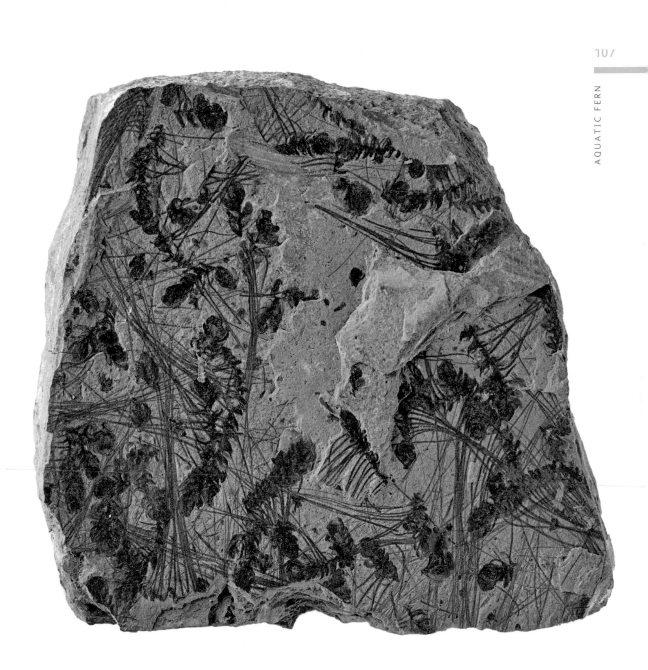

# Savannah and grasslands
## grass phytoliths

Blades of grass can feel abrasive. This property comes from the abundant phytoliths that permeate the tissues of their leaves. The term phytolith comes from Greek, meaning 'plant stone'. These microscopic particles of translucent opal are hard, composed mainly of silica, and quite variable in shape and size. They form naturally when soluble silica is absorbed by the roots and then precipitated in cells and tissues, commonly of the epidermis. The function of phytoliths is debated, but in many species they undoubtedly add an element of support to leaves and stems, and their abrasive nature discourages grazing by herbivores. When plants die and decompose their phytoliths can endure, falling into the soil or getting washed into rivers and lakes. The phytoliths of grasses are especially distinctive. They can be found preserved in fossil soils and in sedimentary rocks where their presence tells a story of the evolution of grass-dominated ecosystems.

Open habitats dominated by grasses, including temperate prairie and tropical savannah, cover up to 40% of Earth's land surface. In geological terms, grasslands are comparatively recent, but they have had a huge influence, not least on the evolution and dispersal of humans out of the savannah of Africa. Nearly half of all grasses have a modified version of photosynthesis called $C_4$, as opposed to the more common $C_3$ photosynthesis. $C_4$ is a sort of biochemical bolt on at the front end of the conventional $C_3$ pathway that makes carbon extraction from carbon dioxide gas more efficient. It gives plants an advantage under conditions of aridity, high temperatures and low soil nitrogen. $C_4$ grasses are mostly tropical and subtropical, whereas $C_3$ grasses are more common in cooler climates. Fossil phytoliths show that open habitat grasses evolved long before grasslands became ecologically prominent. The spread of grasslands began about 25 million years ago, but proceeded at a different pace across continents and at different times for $C_3$- and $C_4$-dominated communities. The shift from closed canopy forest to mixed woodland and grassland ecosystems was driven and sustained by environmental influences, especially those that limit the growth of trees. Increasing aridity and seasonality were important as well as changes in wildfire regimes and the effects of herbivores. Fully open grasslands are a phenomenon of the Late Miocene to Pliocene Epochs and saw the assembly of new communities of large-hoofed herbivores, including horses, rhinos, antelopes and elephants, which evolved to exploit the newly available food resources and open habitats. Humans, too, are creatures of grassland and savannah, and today it is these ecosystems that are most affected and threatened by the way that we use land in agriculture and through increasing urbanization.

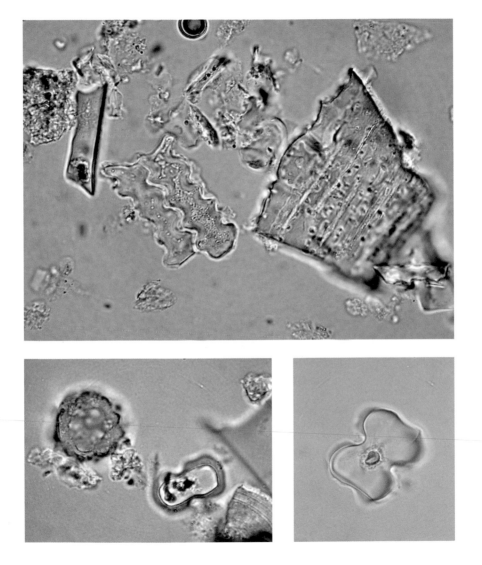

# The taiga
## *Picea banksii*

This fresh-looking cone might have fallen off a spruce tree yesterday, but it is over 3 million years old. It has a woody texture but it is quite friable and crumbles easily between the fingers. The cone was found in a peaty deposit on the north coast of Banks Island in the high Canadian Arctic. Like other fossilized plant remains from this site, it shows no signs of the compression that one would normally expect if it had been buried at depth. The overlapping scaly bracts are open, and some have broken away, indicating that the seeds had been shed long before the cone tumbled to the forest floor. Fossil plants in these sediments were first discovered and collected in 1851 by members of the McClure Arctic Expedition on their voyage in HMS *Investigator* to discover and transit the Northwest Passage, a long searched for shortcut by sea from Europe to Asia. Whilst ashore hunting hare and ptarmigan, members of the crew discovered fossilized logs poking out of the ground. These noteworthy finds made a striking contrast to the treeless tundra of Banks Island, in which the tallest plants are the tiny, creeping arctic willow. The expedition went on to spend a further three tough years trapped in pack ice before finally abandoning ship and making their escape on foot. They were eventually rescued by HMS *Resolute*, returning to Britain in 1854 having successfully completed their transit by a combination of sea travel and sledging. Their fossil discoveries later formed a part of Oswald Heer's *Flora Fossilis Arctica*, which was published in seven volumes between 1868 and 1883. This monographic work was the first comprehensive view of how vegetation evolved in the Arctic through geological time.

The Pliocene Epoch sediments of the Beaufort Formation of Banks Island are well exposed along the steep-walled river valleys of the northwest coast. Here they are composed of flat-lying, loosely consolidated sands, clays and gravels that contain abundant wood and other fossil plants. The flora was dominated by large trees in the pine family, including spruce, pine, hemlock and larch. Smaller broadleaved trees and shrubs included birch and hazel. Grasses were present, and herbs in the carnation and daisy families were especially common. There was a varied fauna of insects, mostly beetles. These high latitude forests were also home to many familiar mammals, including relatives of modern rodents, rabbits, black bears, badgers and beavers. Together, this fossil evidence proves that only 3 million years ago the now desolate coast of Banks Island supported a well-developed boreal forest or taiga.

Today taiga covers a significant part of the northern hemisphere. It accounts for perhaps a third of the total forested area of the planet. Yet, despite its size, it may be the youngest of

OPPOSITE PAGE: Boreal forest or taiga is the world's largest and youngest forested community. Its signature plants are trees in the pine family, most of which are evergreen.

Earth's major forest biomes. Taiga is quite distinct from the forests that clothed the polar regions from 100 to 30 million years ago, when Earth's climate was very much warmer than at present. These earlier polar forests were characterized by deciduous trees and shrubs, incuding varied flowering plants and conifers as well as a broad range of warm-adapted herbs. Modern taiga resembles the Banks Island fossil flora. Its signature plants are trees in the pine family, most of which are evergreen. Redwoods, cypresses and junipers are absent. The origin of the taiga biome is thought to lie in the mountain ranges of North America. Here, there once existed evergreen pine forests. These isolated sites were natural conifer nurseries where many modern species in the pine family evolved. As climate cooling gathered pace, beginning about 23 million years ago, plants migrated south, including the flora of the polar deciduous biome. The mountain forests dominated by pine family evergreens could now expand into lower evelation sites in the high Arctic, and so the taiga was born. The fossil floras of the Beaufort Formation of Banks Island and elsewhere in the Canadian Arctic testify to its recent development during the Pliocene Epoch.

The Pliocene Epoch (5.3 to 2.6 million years ago) is an excellent geological analogue for exploring the likely effects of global warming on Earth systems of the future. During the Pliocene, Earth's continents were in their current positions and species of animals and plants were very similar if not identical to their living relatives. Average global temperature was 3–4°C (37–39°F) warmer than the present day, which is in line with projected estimates of average global warming by the year 2100. However, an increase in Earth's average temperature has a disproportionate effect in the polar regions, where it is amplified. The mean annual temperature of Banks Island during the Pliocene Epoch is estimated to be some 20°C (68°F) warmer, and the tree line would have been shifted much farther north and well into latitudes of extended winter darkness. This warmer world is reflected in the Banks Island flora, which shows that taiga once grew where only treeless tundra is found today. Future global warming driven by human activity might be expected to have a similar effect on the taiga, shifting the whole biome farther north. However, the rapidity of change might outpace the capacity of the taiga to respond. If we returned to the even warmer climates of the older Miocene Epoch (2.3 to 5.3 million years ago), the taiga as we know it today would almost certainly cease to exist. If it could be found anywhere, it would be in the isolated mountain refuges from whence it originally came.

# The origin of flowers
## *Montsechia vidalii*

Bearing a superficial likeness to a sprig of thyme, this little fossils was named *Montsechia* after the Sierra del Montsec, a mountain range located in northern Spain. The rocks of this region are the source of a fine limestone that was used in a method of printing called lithography, and they are also noted for their well-preserved fossil flora and fauna of Early Cretaceous age. Over the years there were many ideas about the nature of *Montsechia*, but recent evidence suggests that it could be a very ancient flowering plant. Specimens were discovered bearing paired fruits, each containing a single minute seed. The preserved features suggested an affinity with flowering plants, but there were no petals or pollen-containing anthers. *Montsechia* seems to be a flowering plant without a proper flower. This apparent contradiction could be resolved if *Montsechia* were a submerged aquatic. In aquatic species today, floating flowers, or blooms carried above the water's surface, can be showy, whereas species that flower underwater mostly have small, very reduced floral organs and no need of petals. Likewise, the male flowers of *Montsechia* might have been highly simplified. Other features supporting an underwater habit include the tiny filamentous leaves, sinuous stems and complete absence of roots in the many hundreds of specimens examined. In many ways *Montsechia* resembles the modern-day *Ceratophyllum*, also known as the hornwort or coontail, a plant that is commonly used to populate aquariums.

No one can say for certain when the first flower bloomed. Flowering plants seem to have originated during the latest part of the Jurassic Period or the earliest part of the Cretaceous Period, but fossils are elusive, inconclusive and frequently controversial. This is part of the mystery of the origin of flowers. Also a problem is that the relationships of flowering plants to the gymnosperms of the Mesozoic Era are uncertain. Knowing the closest kin of flowering plants would provide important clues about their ancestors. A further difficulty is that flower-like structures are a feature of some extinct gymnosperms. At a fundamental level, flowers are just aggregations of seed and pollen-bearing organs, so they can be difficult to tell apart from superficially similar systems in other plants. This has led to the reproductive parts of gymnosperm fossils being mistaken for flowers. In the search for the earliest fossil flowers it is often the case that insufficient features are preserved to be sure about kinship, and some experts would take this view about *Montsechia*. Flowering plants dominate the floras of the modern world, and because of their beauty and fragrance they play special roles in our cultures. It is little wonder that early fossils are highly sought after and prized for what they might be able to tell us about how the flower first evolved.

# Early flowers
## *Silvianthemum suecicum*

Millions of years ago lightning from a thunderstorm ignited a wildfire in woodland that burnt through leaf litter, timber and low-lying shrubbery. Fallen wood and leaves were turned to charcoal. Heavy rains washed the charcoal into a river where it sank into sediments. Today these sediments are quarried for their kaolin, a clay mineral with many industrial uses. Palaeobotanists also seek them out in search of fossil plants. Minute fragments of charcoal can be extracted from the clay simply by soaking it in water and allowing it to disaggregate, then gently sieving to concentrate the plant remains. Fragments of wood and leaf are common, but much rarer and more highly valued are tiny fruits, flowers and even flower buds. Indistinguishable to the naked eye, they can be seen and picked out with the aid of a microscope. This trumpet-shaped piece topped by small flasks is a fossil flower. The ghostly image to the right was made by an intense beam of X-rays generated in a particle accelerator and focused through the fossil, enabling us to peer inside the flower, opening a window onto its internal organs. As if rising from the ashes, floral structures appear from within the charcoal, bringing new perspectives on the early evolution of flowers.

The nature of this fossil flower was worked out from numerous specimens that captured various developmental stages, ranging from buds to mature blooms. The sepals, petals and the anther-bearing stamens are only all preserved together in the bud stage, where they were tightly enclosed and protected. Once the bud opened, the petals and stamens became vulnerable, breaking off during fossilization. The floral bud stage is therefore crucial to interpreting the arrangement of the various parts of the flower. The illustrated specimen shows a later stage, in which all that remains is the female organs. The flasks at the top were the structures that received the pollen, forming a channel to the ovary. Petals and stamens were attached in whorls below these, and the ovary extended into the conical base. Here is where the X-ray image comes in to its own, revealing that the ovary contained hundreds of tiny ovules which, once fertilized, were destined to become seeds. The whole flower is barely 3 mm in length. Since they were first discovered, charcoal flowers have been found at many sites around the world in clays of the Cretaceous Period. Named *Silvianthemum* in honour of Queen Silvia of Sweden, this flower was among the first of its kind to be recognized.

The nature of the ancestral flower is a much discussed topic in botany. A favoured theory likened it to the large flower of modern *Magnolia*. As we learn more about the family history of living species and their fossil record, a different picture emerges. The first flowers were

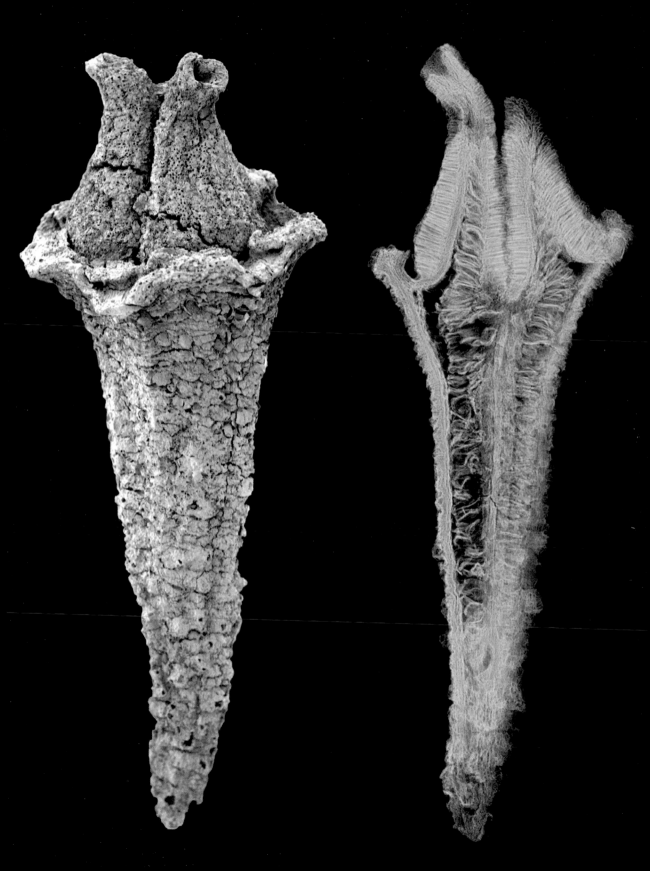

very small, much less than a centimetre in size, and they were carried in groups or clusters. Most were bisexual, but with a simple organization consisting of one or more whorls of basic petal-like parts and another of stamens, all arranged below a small cluster of carpels, each of which contained one or a few ovules. There are no exact modern analogues, but some features are echoed in ancient living lineages. These include botanically iconic but rather obscure plants like *Amborella trichopoda* and groups such as Austrobaileyales, which are shrubs and vines found scattered across Australia, the Pacific Islands, Southeast Asia and the Caribbean. The close living relatives of *Magnolia* have highly varied flowers that range from the large and elaborate in *Magnolia* to the minute and simple flowers of *Sarcandra*, each of which has only a single stamen and carpel. The fossil flowers share some features of these, like bottle-shaped carpels and anthers that split open into valves. Waterlilies were also among the earliest flowering plants, but they too were small-flowered forms. By about 125 million years ago plants with small, simple flowers had started to become noticeable and widespread elements of the world's flora. Larger flowers that would be more familiar to us, including ones in which parts became fused creating structures like floral cups and flowers with strong bilateral symmetry, arose millions of years later.

Today there are more than 370,000 species of flowering plants. One of the main drivers of this astounding variety is thought to be the evolution of pollination mechanisms, and some features of the ancient flowers preserved in charcoal indicate a very early role for insects. These include anthers with specialized modes of pollen release and nectaries. The first pollinators were probably beetles and flies, with pollination by bees, butterflies and moths developing later. Most living species are pollinated by insects and some by birds and mammals, although wind pollination was probably also a strategy for some early flowers and it is favoured under certain circumstances. Today, this strategy predominates in the grasses, where it is an adaption to open habitats and places where animal pollinators are rarer. It is also common in many of the dominant trees of temperate forests, including birches and southern beeches. Pollination happens early in spring before the trees are fully in leaf and benefits from the proximity of individuals and their large populations. These wind-pollinated plants evolved from insect pollinated forebears. Their flowers typically are reduced in size and number of parts, and often they are unisexual with larger numbers of male flowers and a high pollen output. Over the past 100 million years, the flower has proven its versatility time and again.

# Fruits diversify
## *Nypa burtinii*

When discovered in a quarry in Brussels in 1784 this fossil was immediately recognized as a fruit. It is preserved in sand that became cemented with lime and then silicified. The outer husk has partly broken away to reveal the distinctive nut within. Because of its size and appearance, it was first thought to be a fossil coconut. Later, it was recognized to be closer to *Nypa fruticans* or the mangrove palm. These palms are tolerant of brackish water, growing in the mud at the edge of slow-moving tidal rivers and estuaries. They are confined to the tropics, with a natural modern distribution ranging from India to the Pacific Islands. The fruit of the mangrove palm is large, buoyant and dispersed by water. In Europe, it is quite a common fossil in rocks of the Eocene Epoch when tropical climates extended to higher latitudes than they do today.

The fruit is the vehicle for seed dispersal. In the flowering plants, the fruit arise from the female organs of the flower and associated parts, enabling the evolution of many more varied fruit types than in non-flowering species. The *Nypa* palm fruit is technically termed a drupe, meaning that like in a peach or a plum it has a single central stone enclosed in a fleshy coat that does not actively split open to shed the seed or kernel inside. The coat of *Nypa* is waterproof, fibrous and it contains airspaces to aid flotation. The flowering plants have evolved many other ingenious ways to scatter their progeny. One of the most common is by wind. These fruits tend to be small, and they develop wings, plumes or hairs. Another is ballistic dispersal, in which the fruit expels the seeds explosively. Plants have also recruited birds and mammals in various ways. Encouraging ingestion is the most common route for animal dispersal and probably one of the most ancient. These fruits are fleshy and often large and brightly coloured when ripe. Nourishment comes from the fruit coat, and seeds are either discarded or pass through the digestive tract unharmed. Other fruits hitch a ride by developing hooks and spines that attach to fur and feathers. Insects, which are so vital in pollination of the flower, play a lesser role in seed dispersal, with ants being the most important. The small seeds of many herbaceous plants develop an elaiosome, which is a variously shaped fleshy body rich in lipids and proteins that can develop from tissues of the seed or the fruit. Ants take these to their nests to feed their larvae, later discarding the seed. The associations of fruits with their animal dispersers are generally looser and less specific than those of flowers and their pollinators. On the whole, therefore, fruits tend to be somewhat less varied than the flowers that produce them.

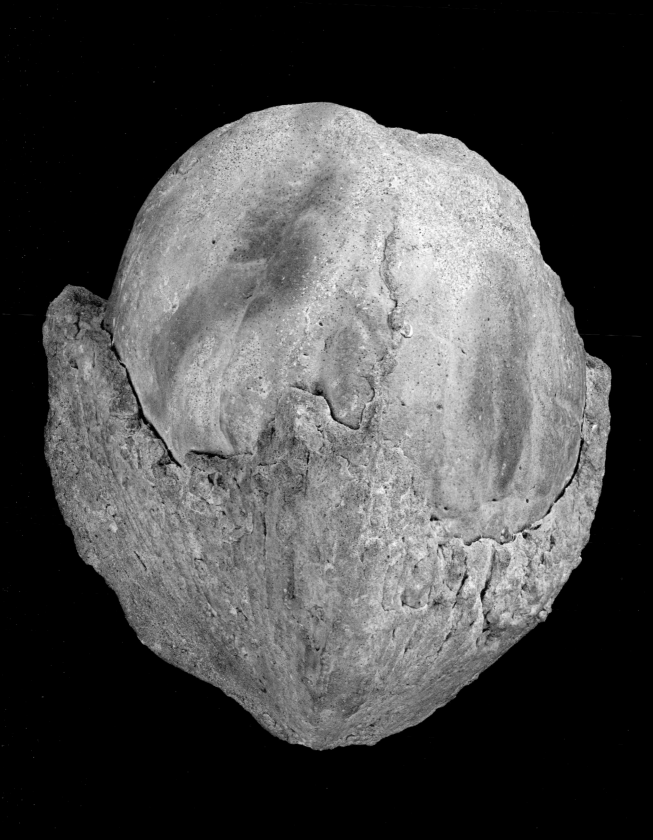

# Sunflowers appear
## *Raiguenrayun cura*

This brush-shaped fossil is the oldest known member of the daisy or sunflower family. Its colour and texture capture something of the feel of Vincent van Gogh's famous paintings of the dying sunflowers, which were built up from thick ochre-coloured brushstrokes invoking the texture of petals and seed heads. The sunflower is distinctive and unusual because what appears to be a single flower actually is a composite of numerous, simple flowers of varied shapes that are grouped together at the end of a stem, collectively acting as a single integrated entity. This compound flowerhead gives the family its name Compositae and also its alternative name Asteraceae, which comes from the classical Latin word *aster*, meaning 'star'. Nearly 10% of the flowers in the world today belong to this family. With 27,400 species, the Asteraceae is second only to the orchids in size, yet measured against the long history of plant evolution, the rise of the sunflowers is a comparatively late phenomenon.

The Asteraceae are thought to originate in southern South America about 70 million years ago, close to the dawn of the Cenozoic Era. In this small corner of the world the new young family survived the catastrophic environmental change that led to the extinction of the dinosaurs, and the earliest stages of their evolution would have taken place in the immediate aftermath, as ecosystems began to recover. Some 20 million years later their numbers and diversity had grown, and they dispersed across the ever-widening southern Atlantic Ocean to Africa, and from there onwards to Europe and Asia, arriving in North America some 30 million years ago. Today species richness and taxonomic diversity is distributed unevenly across continents. South America boasts the highest number of species followed closely by Asia, then North America and Africa. The number of European species is large, but only about a third of that of South America, and Australasia has the fewest. During their early development and expansion Asteraceae became ever more varied but remained comparatively minor components of floras. Their rise to ecological prominence began about 23 million years ago and can be measured by the growing abundance of their pollen preserved in sediments and further rapid evolution of species as seen in their family trees. This later radiation of the sunflowers took place in the Miocene Epoch during a time of climate cooling, growth of ice-sheets in Antarctica and increasing seasonality. This was a period of time that saw the expansion of open, seasonally arid grassland and savannah, which are environments that were also well suited to many species of Asteraceae.

Climate change, along with mountain-building and other geological phenomena, undoubtedly shaped the evolution of Asteraceae over the past 20 million years, but the

*Helianthus annuus*
I Miller's Illustr. Syst. Sext. Syst. Linn. 1777.

family also has characters of its own that are thought to contribute to its success. One of these is its remarkable compound flowerheads. Similar sorts of flowerheads are seen in some other families, yet none is as varied or as highly integrated. In the common sunflower, *Helianthus annuus*, the bright yellow peripheral petals are formed of strap-shaped flowers whereas the yellow or maroon central disc is composed of numerous smaller tubular flowers. The large peripheral flowers are usually the first to open and the last to wither. Furthermore, in the family more broadly the tight arrangement of flowers and the difference in development between peripheral and central parts of the flowerhead create myriad opportunities for varying shape and colour and possibilities for altering patterns of gender within. The beauty of this compound flowerhead is that it concentrates numerous flowers into a single large attractor, so it is no surprise that Asteraceae are mostly insect pollinated, especially by flies and solitary bees. Some of the South American and African species that develop flowering heads composed of long, tubular flowers containing generous supplies of nectar are pollinated by hummingbirds and sunbirds.

The compound flowerhead is also a conspicuous and tempting source of food so, while welcoming pollinators, Asteraceae must fend off herbivores. They have developed various types of armature including tough bracts that protect the flowering head in its bud stage, small dry seeds, woody tissues, and distinctive bristles encircling individual seeds called a pappus. In addition to physical barriers, many species produce bitter and toxic compounds. Some Asteraceae also have an edge in seed dispersal and metabolism. In some species, the pappus of bristles associated with each seed develops into a feathery structure that aids dispersal of the tiny fruits by wind, as in the dandelion. The pappus is thought to have enabled transoceanic dispersal. In other species, stiffer hairs on the fruit enable it to adhere to fur and feathers for animal dispersal. Asteraceae also have an unusual metabolism in which energy is stored in water-soluble polymers of fructose rather than as starch. This is thought to confer an ability to withstand lower temperatures and drought, contributing to the success of the family in seasonally dry and colder environments. The 47 million year old fossil from Argentina is a flowering head composed of many tubular flowers, each about 1.5 cm (½ in) long, with a ring of overlapping bracts near the base all borne on a long stalk. It resembles most closely the lineages of Asteraceae that are today endemic to South America and Africa. The mode of pollination cannot be known for sure in this fossil, but the general shape and size is thought to be consistent with pollination by birds.

# Plants enlist allies
## *Prosopis linearifolia*

This pod enclosing a row of seeds is unmistakable as the simple dry fruit of a legume. Discovered over a hundred years ago, the fossil formed in the sediments of a volcanic lake that is now known as the Florrisant Fossil Beds, Colorado, USA. The words and letters scrawled across the surface and on associated labels tell something of its history, and they also convey the opinions that various people held about which legume it might be. When first prised open, the rock split to create two pieces that were mirror images of each other. These two halves subsequently became separated, beginning journeys that now find them on different continents. One is in the University of Colorado and the other in London's Natural History Museum, which purchased the piece in 1911. Such trading or exchange of fossils was once commonplace but nowadays is increasingly subject to regulation. Over the years this 34 million year old fossil has been attributed to various legumes by different experts, including trees and shrubs in the living genera *Leucaena* and *Prosopis*. Specialists would now say that on its own this little pod is not distinctive enough to place with confidence in a modern genus, but its affinity within the pea or bean family is undisputed.

The recent history of legumes is closely tied to that of human civilization. Legumes have been a staple food for millennia, including among their number peas and beans, soybean, chickpeas, peanuts, alfalfa, carob and liquorice. They are easily recognized by their distinctive fruit, which gives the family its name, yet they are hugely varied and widely distributed. Leguminosae, which is also known as Fabaceae, is the third-largest family of flowering plants, containing an estimated 19,600 species. Most are herbaceous perennials or annuals, but there are many trees, shrubs and vines. One of the most useful qualities of legumes is the ability of many species to fix nitrogen, which is an essential element for plant growth. Nitrogen is naturally available in a usable form in soils and taken up through root systems, but it is required in large amounts and is usually in short supply. Natural sources include nitrates in rocks, the microbial decomposition of dead organisms and animal waste products, but the availability of these varies widely. Legumes have the advantage of being able to make use of nitrogen gas, which is abundant in the atmosphere. In its gaseous form nitrogen is inert and therefore cannot be used directly by plants. In biology, the capacity to convert atmospheric nitrogen into a usable form is restricted almost entirely to the bacteria. Legumes have discovered ways to harness the power of these microbes by allowing them to take up residence inside their roots.

Species of bacteria in the genus *Rhizobia* thrive in soils and they are attracted to the roots of certain legumes. *Rhizobia* enter the plant by the root hair, infecting cells and causing them to proliferate forming growths called nodules. Once inside, the bacteria multiply and begin the process of transforming nitrogen derived from the atmosphere into ammonia. The relationship between plant and bacteria is a mutualistic one. The plant benefits by acquiring a constant supply of nitrogen in a useable form, and in return the bacteria are rewarded with a reliable source of carbon in the form of organic acids. The symbiotic bacteria are not passed on from generation to generation though, so infection needs to take place anew in each legume seedling. The infection process is complex, but it shares similarities with the development of mycorrhizae, which are networks formed between plant roots and fungi. Mycorrhizae evolved long before biological nitrogen fixation in plants, but recent discoveries have found that both processes make use of many of the same genetic mechanisms. During the evolution of nitrogen fixation in legumes, established developmental pathways were co-opted to serve startling new functions, which is a recurring theme in plant evolution.

The time and place of origin of the legumes remains something of a mystery. They probably originated during the Late Cretaceous, but their fossil record is not particularly rich until the Eocene Epoch. At the time that the Florissant Fossil Beds formed, legumes were varied and abundant. One theory traces their origin to the tropics of Africa, from there spreading to other parts of the world while elements of Gondwana were still in close proximity. An alternative theory places them first at high northern latitudes under much warmer climates from where they spread south along various routes into the Americas, Africa and beyond. Today legume trees and shrubs dominate forests of the tropics, especially the lowland rainforests of Africa and South America but also seasonally dry forests. In contrast, most of the temperate species are herbaceous perennials or annuals. The secrets of their undoubted success are debated, but their capacity for nitrogen fixation probably played a key role. Legumes have a high nitrogen lifestyle with leaves and seeds that are exceptionally rich in proteins. This enables them to develop rapidly and to sustain high photosynthetic rates, making them unusually flexible and responsive to environmental change at a local scale. Many also have defences against herbivory that are based on nitrogen-rich chemicals. These ecological strategies gave legumes an edge, and their symbiosis with *Rhizobia* enables them to feed their nitrogen-demanding habits.

# Dispersing fruits
## assorted fossil fruits

These assorted fruits were caught in a fast-moving surge of extremely hot volcanic gas, ash and rock known as a pyroclastic flow. The ash formed moulds of their outer surfaces and the heat destroyed their internal tissues creating voids that were quickly filled with minerals. It is these casts that we see, faithfully preserving the shapes of the fruits but none of their internal tissues. Many of the specimens are distinctive enough to attribute to a family or even to a genus of plants. The largest is about the size of a lemon. Its shape tells us that it belongs to *Sterculia*, which is a genus of trees and shrubs colloquially known as tropical chestnuts. The genus was named after Sterculius of Roman mythology, who was the god of manure, a reference to the unpleasant aroma of the flowers. These deposits also preserve a rich record of fossil primates, including species close to the ancestors of great apes and gibbons.

Rusinga Island lies at the mouth of the Gulf of Winam in Lake Victoria, one of the African Great Lakes. Its rich fossil flora provides us with an insight into the environments occupied by our primate ancestors. The animals and plants in this part of Africa, 18 million years ago, lived in the shadow of a large volcano that erupted periodically, depositing massive volumes of ash, cobbles and boulders in the vicinity. Today the eroded remains of the Kisingiri volcano are still visible as a partial ring of high ridges and a central dome-shaped hill on the mainland overlooking Rusinga Island. Between eruptive phases, vegetation regrew developing into open woodland. In some rock layers preserved tree stumps are visible, enabling their size and density to be measured, proving that an extensive closed canopy forest also developed from time to time. The climate was tropical but seasonally dry, supporting a rich vegetation of trees, shrubs and climbers with an herbaceous understorey. This was the setting for a varied fauna including extinct primates that tell a story of the early evolution of apes in East Africa. One of the best known is *Proconsul*, which possessed both monkey and ape-like qualities. Like monkeys, *Proconsul* probably had a horizontal posture, walking and resting on its hands and feet, but several of its features suggest more ape-like climbing and clambering behaviours. In particular, the shapes of its toes indicate powerful grasping capabilities and forelimb anatomy shows mobility of shoulder and elbow and a capacity to rotate the wrist. Like apes, *Proconsul* also lacked a tail. The combination of forelimb anatomy and powerful grasping would have enabled it to climb trees and to live an arboreal lifestyle in the dense, tropical forest that flourished in the rich soils on

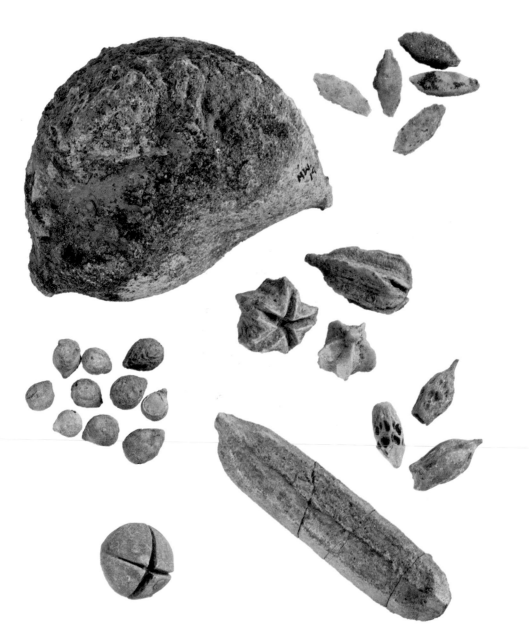

'Loquats.
Calcutta.
Mar.

The
    Nut is like a milky
Filbert in flavour. and Consistency.

A curious
Calcutta New Market Fruit from
Singapore, called The Chinese Almond.  Feb.

the plains around the flanks of the volcano. Most primates eat fruits, and many eat them in large quantities. The arboreal lifestyle suits this dietary preference enabling them to seek out and to reach a valued resource. Most primates also have another advantage over other mammals that helps them to forage fruits. Like us they have trichromatic vision, with three distinct types of colour receptors called cone cells in the retina of the eye. These are sensitive to different wavelengths of light, defining the range of the visible spectrum and the palette of colours that we are able to discern. Other mammals are typically dichromats, meaning that they have one fewer type of colour receptor, with the consequence that they are unable to separate green and red. One theory about the evolution of trichromacy in primates is that it arose as an adaptation for finding fruit. Ripe fruits are a nutritious but scarce resource often available only on a seasonal basis. The ripening process is usually accompanied by a change in colour. Objects that differ in colour from the background catch the eye, but only if you can see the difference. Trichromatic vision enhances the ability of primates to spot coloured fruits among the brown twigs and the green leaves of a tree.

One of the most common adaptations of the flowering plants is to produce edible fruits for seed dispersal and then to advertise them to animals when the time is right by their odours and colours. This set in motion a loose co-evolution between fruits and seeds and their animal dispersal agents that began over 80 million years ago, but seems to have accelerated and diversified some 30 million years later from the middle of the Eocene Epoch. What changed at this time was that many of the important modern groups of fruit-eating animals began to diversify, including the primates and the fruit-eating birds and bats. The evolution of fleshy fruits in varied families of flowering plants became increasingly common during the Miocene Epoch, as is evident in the Rusinga Island flora. Today plants with fleshy fruits account for about 40% of woody species in temperate forests and well over 70% in tropical forests. To the plant, the animal is just a handy means of dispersal. Plants have become ever more adept at exploiting the needs, harnessing the behaviours, and shaping the evolution of unwitting birds and mammals in their quest to disperse their own progeny. At the same time they have added a welcome splash of colour and many new textures and flavours to our menu.

# Humans follow flora
## *Pinus*

In Norfolk, on the east coast of England, low cliffs formed of Pleistocene sediments are rapidly eroding into the North Sea. Here and there one can see old sea walls and other structures that were put in place over the years in failed attempts to stem the tide. As the sediments are washed away, plant and animal fossils as well as artefacts are exposed that shed light on a very early period of human prehistory. Evidence for the presence of humans often comes only from the things that we leave behind, which in this case were sharp-edged flint tools. More recently fossilized footprints were found marking an ancient path taken by a small group of juveniles and adults, possibly a family. The artefacts and footprints were in sediments that were originally deposited near the mouth of the ancestral River Thames over 850,000 years ago. At that time the river followed a very different course, discharging into the North Sea some 150 km (93 miles) north of where it does today. The pine cone pictured is one element of a fossil flora that provides ecological context showing how, following dispersal from Africa, some of our ancestors began to adapt to the challenges of life in northern climes.

Since the tree that produced this pine cone grew, ice-sheets have developed over vast swathes of Europe, Asia and North America and retreated at least eight times. During each period the flora of the British Isles was nearly completely erased by glaciers only to re-establish again when the ice retreated. Where flora goes, fauna follows. Human populations are thought to have ebbed and flowed in tune with these glacial cycles, surviving in southern sanctuaries during cold stages and moving north again as the climate warmed. This pine cone is from the most northerly uncontested Early Pleistocene archaeological site. Together with other fossil plants and pollen it tells of a relatively warm period between glacial cycles when the ancestral River Thames flowed through grasslands and extensive conifer-dominated woodland. Mammals found here include red deer and various extinct species of horse, hyena, elk and mammoth. The environment and climate resembled that of southern Scandinavia today near the transition between the temperate and boreal zones. Occupation of these northern sites would have presented novel challenges to hunter-gatherers from the south. Although summer would have provided plentiful food resources, inhabitants would have been challenged by sparse mammalian prey, short winter days for foraging, and severe winter cold. As our ancestors moved north out of Africa and the Mediterranean Basin they were not simply tracking familiar southern habitats but had to be resourceful at adapting to the new challenges of life on the edge of the boreal zone.

# Domestication of grain
## *Triticum aestivum*

These charred grains of wheat were found interred within a Roman-period grave in southern Britain. The charring happened on a funeral pyre, which inadvertently rendered the grains more resistant to decomposition. The ritual inclusion of foodstuffs in ancient burials was quite common, carrying social and cultural significance. The practice might have originated with early indigenous religions of the Middle East, in which wheat played a significant role, being associated with various deities. Bread continues to feature in worship and ritual in the Jewish, Christian and Muslim traditions. Wheat is the pre-eminent food crop in temperate regions of the world. It owes its success to its adaptability, high potential yield, its capacity to be harvested mechanically and excellent storage properties. Perhaps even more significant is the gluten it contains, which is both elastic and viscous when deformed, allowing doughs to form and to be processed into products as varied as breads, pastas and noodles. Wheat also contributes essential amino acids, minerals, vitamins, beneficial phytochemicals and fibre to the human diet, and these are particularly enriched in whole-grain products. The domestication of wheat and the evolution of human societies in much of the Old World is so intimately entwined, it is little wonder that we cannot bear to part company, even in the grave.

Nearly 70% of calories consumed by humans come from 15 species of plants, the top four of which are rice, wheat, maize and sugarcane, all of which are grasses. Few people have ever seen or would recognize the unpromising wild species that are the progenitors of these remarkably productive crops. What sets them apart is a common suite of features known as the domestication syndrome. Depending on the species, these can include loss of toxic or unpleasant-tasting compounds, modifications to prevent disintegration of the seed-bearing head and scattering of seed before harvest, simultaneous ripening and increased seed size, the ability to loosen the grain by threshing, loss of seed dormancy, and many other characteristics that make them easier to farm and to process. The first farmers selected the earliest cultivated forms of wheat from wild populations, presumably because of their superior yield and other desirable characteristics. This happened over 10,000 years ago during the Neolithic, which saw a transition from hunter-gathering of food to settled agriculture. These earliest cultivated forms of wheat originated in southeastern Turkey, entering Europe via Greece about 8,000 years ago. From there they moved westwards across to Italy, France and Spain and northwards through the Balkans to the Danube, finally reaching the UK and Scandinavia about 5,000 years ago. Wheat spread east into

Gramineae (Hordeae.)

Triticum vulgare L.

WMüller n.d.Nat

Asia by way of Iran reaching China about 3,000 years ago and to Africa, initially via Egypt. It was introduced by the Spanish to Mexico in 1529 and by the British to Australia in 1788. Today wheat is one of the most productive cereal crops, exceeded only by maize in tonnage produced annually. It is unrivaled in its range of cultivation, more widely grown than any other staple crop, with a total area under cultivation equivalent to the size of Greenland.

The varieties of wheat and their properties as food crops are as much a product of their ancient history of natural hybridization as the long selection for desirable properties by farmers. Wheat genetics is more complicated than that of most other domesticated species. Some wheats are diploid, with two sets of chromosomes, but many are stable polyploids with four (tetraploid) or six (hexaploid) sets of chromosomes. Among the first plants to be domesticated was einkorn wheat, which is diploid. Today einkorn is considered a relict crop that has largely been supplanted by agronomically superior wheats. The natural hybridization of two wild grasses some 500,000 years ago gave rise to emmer wheat, which is a tetraploid . Cultivated emmer and durum wheats are derived from wild emmer. Durum is adapted to the dry Mediterranean environment where it is used to make semolina and pasta. About 8,000 years ago an emmer or durum variety under cultivation hybridized with yet another wild grass producing a hexaploid. This might have happened more than once, giving rise to the varieties that we call spelt and bread wheat. Today bread wheat is by far the most commonly grown accounting for 95% of wheat production.

Humans have shaped the evolution of wheat by selecting and breeding from populations with desirable properties over thousands of years. At first this happened unconsciously by farmers preserving the most valued individuals; later breeding became more systematic, working towards pre-determined goals that have extended the crop's geographic range and its yields. Most recent growth in wheat production has occurred through gains in yield rather than expansion of harvested land, and bread wheat is now so domesticated that it is unable to survive in the wild. Breeders continue to address the challenges of disease resistance, increasing yield whilst maintaining quality and reducing dependence on nitrogen fertilizers. Today we have the technology to manipulate the wheat genome directly to enhance desired traits, promising more rapid and targeted developments, bringing a new phase of our relationship with wheat that is proving controversial. The wheat grain has been treated with reverence by humans for millennia, and it remains to be seen whether consumers will eventually come to accept genetically modified varieties.

# Wetlands for well-being
## *Phragmites australis*

These tubes, with their distinctive longitudinal ribs and perpendicular nodes, are parts of the underground rhizome and stem of a fossilized reed preserved in silica. They were found in sediments of the Faiyum Basin, just north of the ruins of Dimeh es-Seba, southwest of Cairo, Egypt, believed to have been founded by Pharaoh Ptolemy II in the third century BC. Dimeh es-Seba was once a thriving religious centre supported by a farming economy, but it now stands 2.5 km (1½ miles) from the shore of present-day Lake Qarun, surrounded by barren desert. Like the town, these reeds would once have thrived at the margins of a lake, and both bear witness to a time when wetlands were widespread in the region. Greater areas of wetlands have existed in the Faiyum Basin since Pleistocene times, fed by waters from the nearby river Nile. Their extent fluctuated over time, at first regulated naturally by changing input from Nile floods and later managed by pharaohs of the Middle Kingdom (2050–1710 BC) who launched an ambitious construction programme of canals and dams. Until at least the mid-twentieth century, wetlands were widely perceived as swamps suitable only for draining and filling for development or control of mosquito-borne diseases. Today they are recognized as among the most productive yet highly threatened ecosystems on Earth.

Reeds are one of the most readily recognizable and widely distributed wetland plants in the world, forming extensive beds on the shores of lakes and gulfs, along riverbanks and in peatlands. They have served human societies well since ancient times, where they have been used as fodder and readily available raw materials for the building of dwellings and the crafting of furnishings. Reedbeds cleanse our environment by purifying water that passes through, and increasingly they are seen as a potential source of renewable energy. Wetlands contribute to human well-being in many ways, providing food, including rice and fish, drinking water, fibre and fuel. They influence climate, help to alleviate pollution and lessen the risk of catastrophic flooding. They offer recreational possibilities and tourism benefits. The health risks of wetlands can be minimized through education and the management of waterborne diseases. Wetlands are frequently sites of pilgrimage and spiritual fulfilment, and their waters are used in rituals and for healing purposes. Ancient Egyptians considered the Nile to be a gift of the gods and they equated the river with life itself. Wetlands and their botanical benefits are threatened by population growth, which brings demands for infrastructure, land conversion and water for drinking and irrigation. These fossils show that when wetlands diminish one consequence is environmental degradation. Over parts of the Faiyum, their loss has led to dusty ruins in a parched landscape.

# Saving the seeds
## *Silene stenophylla*

In northeastern Siberia, icy deposits of wind-blown dust and sand conceal the bones of large mammals from the latter part of the Pleistocene Epoch, including mammoth, woolly rhinoceros, bison, horse and deer. Also conserved, now deep below the present-day surface, are ancient burrows dug by Arctic ground squirrels. These contain the remains of frozen plants that were harvested by the male of the species during late summer, so that when it awoke from hibernation in the spring it would have a ready store of food to tide it over until warmer weather stimulated new growth in the hardy flora of the treeless tundra. These little larders were never fully emptied and numerous seeds remain, testifying to the squirrels' tastes for sedges (*Carex* spp.), alpine bearberry (*Arctous alpina*), arctic dock (*Rumex arcticus*) and the narrow-leafed campion (*Silene stenophylla*). Radiocarbon analysis dates the seeds to more than 31,000 years old, yet when examined under laboratory conditions many still showed some residual physiological activity in their cells and tissues. None was able to germinate on its own, but when tissues connecting seed to ovary wall were carefully dissected out and nurtured on a nutrient medium, plants of the narrow-leafed campion began to take form. Once these were transferred to soil they grew normally, developing flowers, fruits and setting seed. A flicker of life was preserved within these deep-frozen seeds, which are the most ancient flowering plants to be regenerated, true relicts of the Pleistocene Epoch.

Seeds are tiny, embryonic plants enclosed in a protective covering. In many species they are resilient, and once dispersed can remain viable in soils for years. These tiny time travellers contribute to soil seed banks, which play important ecological roles in the rapid regeneration of vegetation following disturbance or natural catastrophe. The longevity of seeds in soils is extremely variable, being influenced by many factors, including the intrinsic physiology of the seed, ambient temperature and humidity, and predation by fungi and animals. Few seeds remain viable for more than 100 years, yet there are several known instances of exceptional longevity. The oldest seed to germinate was a 2,000-year-old date palm, *Phoenix dactylifera*. Some seeds were discovered in the 1960s beneath rubble in the historic fortress of Masada, built by Herod the Great on the edge of the Judean Desert between 37 and 31 BC. One that was planted in 2005 grew into a healthy seedling. Another example of great longevity is the sacred lotus, *Nelumbo nucifera*. Seeds collected from a dry lake in Liaoning Province, China, ranged in age from 200 years to 1,300 years. Many were successfully germinated, but some

BELOW: Seed of the modern red campion, *Silene dioica*, highly magnified resembles the seeds of the narrow leaved campion, *Silene stenophylla*, found in Pleistocene Epoch sediments of Siberia.

OPPOSITE PAGE: *Silene*, also known as the campion or catchfly, is a genus of about 700 species widely distributed in temperate and alpine habitats especially of the northern hemisphere.

plants exhibited abnormalities that were attributed to hundreds of years of exposure to background gamma radiation in the soil. If seeds are able to survive for hundreds of years and even for millennia under less than ideal conditions, stored with care they provide a means of conserving plant diversity long into the future.

Plants are fundamental to life on Earth and a decline in their numbers is detrimental to our quality of life. It is estimated that as many as 20% of plant species are threatened with extinction. The biome with the most endangered species is tropical rainforest, where the greatest threat to plants lies in the conversion of forest to agricultural land and the unsustainable harvesting of natural resources. Conserving whole communities is recognized as a paramount strategy for maintaining the evolutionary potential of plant life, but the rate of environmental change is so rapid that it is not always possible to conserve plants in the wild. So valuable is our plant genetic diversity that seeds are now collected and preserved away from their natural habitats in seed banks.

Saving the seeds as an approach to conservation was originally practised to retain diversity among crop species, and especially to conserve historic varieties that are no longer widely cultivated. The largest collection of crops preserved in this way is stored in the Svalbard Global Seed Vault, which lies deep inside a mountain on the Norwegian Arctic island of Spitsbergen. This vast depository is a back-up for the major international agricultural seed banks that supply most of the seeds used to breed new crop varieties. Over the past two decades an increasing number of botanic gardens and other institutions have established seed banks to conserve wild plants, including species that are related to domesticated crops. The largest and most diverse collection now resides in the Millennium Seed Bank of the Royal Botanic Gardens, Kew, located near London. Seed banking involves collecting fresh seeds, drying and storing them in a frozen state at -20°C (-4°F). While this process works for many seeds, it is not suited to about a third of species. These require special treatment called cryopreservation, which involves removing the embryo and freezing it at the much colder temperature of -196°C (-320°F). Like the Pleistocene seeds from the Siberian permafrost, these frozen assets can be restored at a later date. Preserving in this way does not stave off environmental change or lessen the impact of losing forests, savannah and wetlands that have taken eons to evolve. Where it can help is to save species from extinction. By banking our botanical capital now, we are able to conserve plant genetic diversity, which is an investment in our children's futures.

p.4
*Populus latior* leaf
Miocene; Öhningen,
   Baden-Württemberg,
   Germany
NHMUK V 18
Leaf width 11 cm (4 in)

p.7
Dendritic pyrolusite
   pseudofossil
Late Jurassic; Solnhofen,
   Bavaria, Germany
NHMUK BM.39897
Rock width 14.5 cm (5¾ in)

p.9
Banded Iron Formation
   rock
Paleoproterozoic;
   Hamersley Basin, Pilbara,
   Western Australia
NHMUK AQ-PEG-2016-33
Width at base 2.1 m (6¾ ft)

p.11
*Bangiomorpha pubescens*
   filament
Mesoproterozoic; Somerset
   Island, Arctic Archipelago,
   Canada
Harvard University
   Paleobotanical Collections
   63008
Filaments diameter up to
   30 μm

p.13
*Coelosphaeridium*
   calcareous algae
Ordovician; Ringsaker,
   Hedmark, Norway
NHMUK V 28247
Rock width 6 cm (2¼ in)

p.17
*Cooksonia pertoni* spore
   capsules
Silurian; Herefordshire,
   England, UK
NHMUK V 58010
Fossil length 1 cm (¼ in)

p.21
Charcoal xylem cell
Lower Devonian;
   Shropshire, England, UK
NMW 99.20G.1
Cell diameter 20 μm

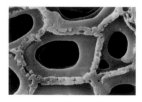

p.23
*Thursophyton elberfeldense*
   stems
Middle Devonian; Elberfeld,
   Wuppertal, Germany
NHMUK V 17195
Rock width 24 cm (9½ in)

p.25
Megaspores in brown coal
Carboniferous,
   Mississippian; Pobiedenko,
   Moscow, Russia
NHMUK V13068
Spore diameters 0.5 mm

p.29
*Xenotheca devonica* seed-
    bearing cupules
Upper Devonian; Devon,
    England, UK
NHMUK V 31136
Fossil length 2.5 cm (1 in)

p. 31
*Rhynia gwynne-vaughanii*
    cross-section of stem
Lower Devonian; Rhynie,
    Aberdeenshire, Scotland, UK
NHMUK Scott Collection
    3133
Stem diameter 2 mm

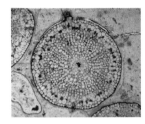

p.33
*Stigmaria ficoides* root
Carboniferous,
    Pennsylvanian;
    Clackmannanshire,
    Scotland, UK
NHMUK V 3111
Rock width 25 cm (9¾ in)

p.35
*Eospermatopteris* tree stump
Middle Devonian; Gilboa,
    New York State, USA
Specimen on display
    roadside Gilboa
Width at base 1 m (3 ft)

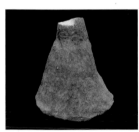

p.39
*Lepidodendron subdichotum*
    branch
Carboniferous,
    Pennsylvanian; locality
    unknown, UK
NHMUK 39031
Width at base 6.5 cm (2½ in)

p.43
*Psaronius brasiliensis* trunk
Lower Permian; Rio Grande
    do Sul, Brazil
NHMUK V 9002a
Diameter 18 cm (7 in)

p.47
*Agathoxylon*
    (*Araucarioxylon*) trunk
Upper Triassic; Petrified
    Forest National Park,
    Arizona, USA
NHMUK V 28224
Diameter 25 cm (9¾ in)

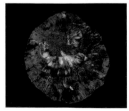

p. 51
Fossil tree stumps
Eocene; Axel Heiberg
    Island, Qikiqtaaluk
    Region, Nunavut, Canada
Stump diameter exceeds
    1 m (3 ft)

p.53
*Retesporangicus lyonii* spore
    capsule
Lower Devonian; Rhynie,
    Aberdeenshire, Scotland,
    UK
School of Geosciences,
    University of Aberdeen,
    thin section 149-CT-B
Flask width at base 50 μm

p. 55
Conifer Wood
Upper Jurassic; Isle of
Portland, Dorset, England,
UK
NHMUK V 68808
Diameter 26 cm (10¼ in)

p.57
*Trigonocarpus parkinsonii*
seeds
Carboniferous,
Pennsylvanian; Locality
Unknown, UK
NHMUK Bowerbank
Collection 41155
Seed length 2 cm (¾ in)

p.59
*Pegoscapus cf. peritus* fig
wasp with pollen
Miocene; Dominican
Republic
NHMUK I.II.3039
Body length 1 mm

p.61
Coprolites (fossilized faecal
pellets)
Middle Jurassic; Roseberry
Topping, Yorkshire,
England, UK
NHMUK V 58510
Rock width 8 cm (3 in)

p.63
*Stigmaria* trunk with
attached root system
Carboniferous,
Pennsylvanian; Brymbo,
Clwyd, Wales, UK
NMW 2015.4G.1
Length of vertical trunk
1.7 m (5½ ft)

p.65
*Cladophlebis australis* leaf
Middle Jurassic; Beaudesert,
Queensland, Australia
NHMUK V 24557
Rock width 8 cm (3 in)

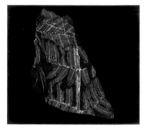

p. 67
*Asterophyllites* leaves and
*Palaeostachya wagneri* cones
Carboniferous,
Pennsylvanian; Hirwaun,
Mid Glamorgan, Wales, UK
NHMUK Taylor Collection
V 68610
Rock width 22 cm (8½ in)

p.71
*Bucklandia anomala* trunk
Early Cretaceous, Cuckfield,
West Sussex, England, UK
NHMUK V 3690
Length 12 cm (4¾ in)

p.75
*Araucaria mirabilis* cone
Middle Jurassic; Cerro
Cuadrado Petrified Forest,
Patagonia, Argentina
NHMUK V 58403
Width 6 cm (2¼ in)

p.77
*Archaeopteris hibernica*
branch with leaves and
cones
Upper Devonian; Kiltorkan,
Kilkenny, Ireland
NHMUK unregistered
Rock length 68 cm (26¾ in)

p.79
*Physostoma elegans* seed
Carboniferous,
Pennsylvanian; Rochdale,
Greater Manchester,
England, UK
NHMUK Oliver Collection
1683
Oval body length 2.5 mm

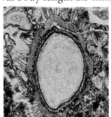

p.81
*Ginkgo cranei* leaf
Paleocene; Almont, North
Dakota, USA
NHMUK V 68763
Leaf width 7 cm (2¾ in)

p. 85
*Agathis jurassica* leaf with
fish
Upper Jurassic; Talbragar,
New South Wales,
Australia
NHMUK P 12440
Leaf length 14 cm (5½ in)

p.89
*Monanthesia saxbyana*
trunk
Lower Cretaceous; Isle of
Wight, England, UK
NHMUK V 63589
Width field of view 12 cm
(4¾ in)

p.93
*Pagiophyllum peregrinum*
twig
Lower Jurassic; Dorset,
England, UK
NHMUK V 68809
Branch length 7 cm (2¾ in)

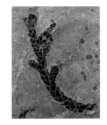

p.95
Various pyritized fruits
Eocene; Isle of Sheppey, Kent,
England, UK
Clockwise from top left:
NHMUK V 64885, NHMUK
V 64872, NHMUK V 64944,
NHMUK V 64922, NHMUK
V 64938. Longest 2 cm (¾ in)

p. 97
*Acer trilobatum* leaf
Miocene; Öhningen,
Baden-Württemberg,
Germany
NHMUK V 18429
Leaf width 9 cm (3½ in)

p.99
*Nothofagus beardmorensis*
leaves
Neogene; Beardmore
Glacier, Transantarctic
Mountains, Antarctica
Leaf width 3 cm (1 cm)

p.101
*Ficus* leaves
Pleistocene; Kharga Oasis
  Depression, Western
  Desert, Egypt
NHMUK V 27751
Rock width 23 cm (9 in)

p. 103
*Glossopteris indica* and
  *Glossopteris stricta* leaves
Upper Permian; Sillewada,
  Nagpur, Maharashtra,
  India
NHMUK V 64045
Rock length 60 cm (23½ in)

p.105
Leaf cuticle of *Ginkgo
  huttonii* with stomata
Middle Jurassic; Whitby,
  Yorkshire, England, UK
NHMUK V 27499b
Width of field of view 1 mm

p.107
*Azolla* whole plants
Eocene; Driftwood
  Creek, Smithers, British
  Columbia, Canada
NHMUK V 55499
Rock width 5 cm (2 in)

p.109
Various grass phytoliths
  from savannah
Miocene; Nebraska, USA
Mostly less than 20 μm

p.111
*Picea banksia* cone
Pliocene; Banks Island,
  Northwest Territories,
  Canada
NHMUK V 57063
Length 6 cm (2¼ in)

p.115
*Montsechia vidalii* leafy
  branch
Lower Cretaceous; Lérida,
  Catalonia, Spain
NHMUK V 32292
Length 5 cm (2 in)

p. 117
*Silvianthemum suecicum*
  flower
Late Cretaceous; Åsen,
  Dalarna County, Sweden
Swedish Museum of Natural
  History S171578
Length 3.3 mm

p. 121
*Nypa burtinii* fruit
Eocene; Schaerbeek,
  Brussels, Belgium
NHMUK V 21762
Width 11 cm (4¼ in)

p.123
*Raiguenrayun cura* flowers
Eocene; Estancia Don
  Hipólito, Río Negro
  Province, Argentina
Museo del Lago Gutiérrez
  Dr. Rosendo Pascual de
  Geología y Paleontología
  MLG 1156
Width of flower head 3 cm
  (1 in)

p.127
*Prosopis linearifolia* seed
  pod
Eocene; Florissant Fossil
  Beds, Colorado, USA
NHMUK V 12384
Pod length 6 cm (2¼ in)

p.131
Various Fossil Fruits
Miocene; Rusinga Island,
  Lake Victoria, Kenya
NHMUK unregistered
Length of largest 10 cm
  (4 in)

p.135
*Pinus* cone
Pleistocene; Happisburgh,
  Norfolk, England, UK
NHMUK V 68778
Width 1.75 cm (¾ in)

p.137
*Triticum aestivum* grain
43 to 410 AD; Wiltshire,
  England, UK
NHMUK V 6622
Grain length 5 mm

p.141
*Phragmites australis* stem
Pleistocene-Holocene;
  Faiyum, Western Desert,
  Egypt
NHMUK V 16854
Length longest 11 cm (4¼ in)

# The Earth over time

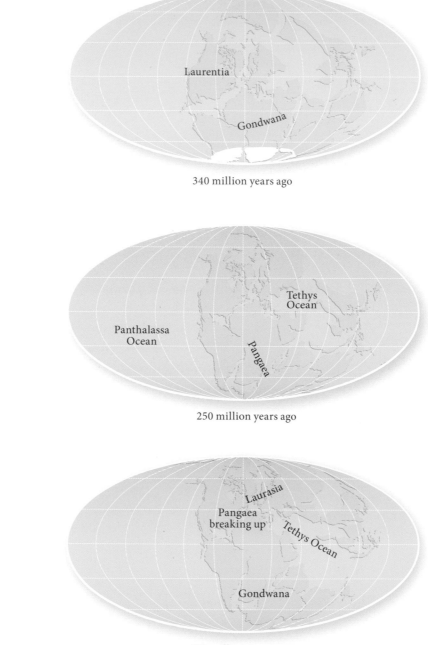

340 million years ago

250 million years ago

170 million years ago

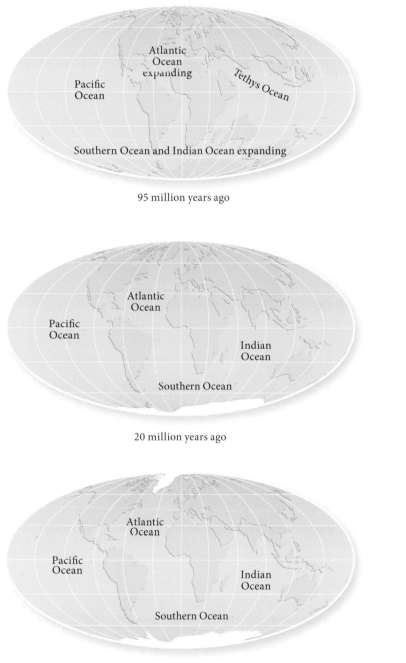

95 million years ago

20 million years ago

2.5 million years ago

# Geological timescale

| EON | ERA | PERIOD OR EPOCH | | AGE (millions of years) |
|---|---|---|---|---|
| PHANEROZOIC | CENOZOIC | NEOGENE | Holocene | 0.012 |
| | | | Pleistocene | 2.6 |
| | | | Pliocene | 5.3 |
| | | | Miocene | 23 |
| | | | Oligocene | 34 |
| | | PALAEOGENE | Eocene | 56 |
| | | | Paleocene | 66 |
| | MESOZOIC | | Cretaceous | 145 |
| | | | Jurassic | 201 |
| | | | Triassic | 252 |
| | PALAEOZOIC | | Permian | 299 |
| | | | Carboniferous | 359 |
| | | | Devonian | 419 |
| | | | Silurian | 443 |
| | | | Ordovician | 485 |
| | | | Cambrian | 541 |
| PRECAMBRIAN | | | Ediacaran | 635 |
| | | | Cryogenian | 850 |
| | | | | 4600 |

1. The vertical (time) axis is not to scale
2. Only the two youngest periods and none of the eras of the Precambrian are shown
3. Epochs rather than periods are specified for the Cenozoic

# Index

# Further information

## BOOKS

Ash, Sidney. *Petrified Forest: A Story in Stone.* 2nd ed. Arizona: Petrified Forest Museum Association, 2005.

Beerling, David J. *Making Eden: How Plants Transformed a Barren Planet.* Oxford: Oxford University Press, 2019.

Cantrill, David J., and Imogen Poole. *The Vegetation of Antarctica through Geological Time.* Cambridge: Cambridge University Press, 2012.

Cleal, Christopher J., and Barry A. Thomas. *Introduction to Plant Fossils.* 2nd ed. Cambridge: Cambridge University Press, 2019.

Crane, Peter R. *Ginkgo: The Tree That Time Forgot.* New Haven: Yale University Press, 2013.

Crawford, Robert MacGregor Martyn. *Tundra-Taiga Biology: Human, Plant, and Animal Survival in the Arctic.* Oxford: Oxford University Press, 2014.

Dörken, Veit Martin, and Hubertus Nimsch. *Morphology and Identification of the World's Conifer Genera.* Remagen, Germany: Kessel, 2019.

Falkowski, Paul G., and Andrew H. Knoll, eds. *Evolution of Primary Producers in the Sea.* Amsterdam; Boston: Elsevier Academic Press, 2007.

Friis, Else-Marie, Peter R. Crane, and Kaj R. Pedersen. *Early Flowers and Angiosperm Evolution.* Cambridge: Cambridge University Press, 2011.

Graham, Alan. *Land Bridges: Ancient Environments, Plant Migrations, and New World Connections.* Chicago; London: University of Chicago Press, 2018.

Hanson, Thor. *The Triumph of Seeds: How Grains, Nuts, Kernels, Pulses, & Pips, Conquered the Plant Kingdom and Shaped Human History.* New York: Basic Books, 2015.

Hill, Robert S., ed. *History of the Australian Vegetation: Cretaceous to Recent.* Cambridge: Cambridge University Press, 1994.

Laws, Bill. *Fifty Plants That Changed the Course of History.* Newton Abbot, Devon: David & Charles, 2010.

Meyer, Herbert. W. *The Fossils of Florissant.* Washington: Smithsonian Books, 2003.

Mitsch, William J, and James G Gosselink. *Wetlands.* 5th ed. Hoboken, New Jersey: John Wiley & Sons, 2015.

Morley, Robert J. *Origin and Evolution of Tropical Rain Forests.* Chichester: John Wiley & Sons, 2000.

Scott, Andrew C. *Burning Planet: The Story of Fire through Time.* Oxford: Oxford University Press, 2018.

Stokland, Jogeir N, Juha Siitonen, and Bengt Gunnar Jonsson. *Biodiversity in Dead Wood, Ecology, Biodiversity and Conservation.* Cambridge: Cambridge University Press, 2012.

Taylor, Paul D., ed. *Extinctions in the History of Life.* Cambridge: Cambridge University Press, 2004.

Taylor, Paul D., and Aaron O'Dea. *A History of Life in 100 Fossils.* London: Natural History Museum, 2015.

Taylor, Thomas N., Edith L. Taylor, and Michael Krings. *Paleobotany: The Biology and Evolution of Fossil Plants.* 2nd ed. Amsterdam; Boston: Academic Press, 2009.

Wilsey, Brian J. *The Biology of Grasslands.* New York: Oxford University Press, 2018.

## WEBSITES

Botanic Gardens Conservation International
https://www.bgci.org/

Botanical Society of Britain & Ireland
https://bsbi.org/

Botanical Society of America
https://cms.botany.org/home.html

Global Strategy for Plant Conservation 2011-2020
https://www.cbd.int/gspc/

International Organisation of Palaeobotany
https://palaeobotany.org/

Linnean Society of London
https://www.linnean.org/

Millennium Seed Bank, Royal Botanic Gardens Kew
https://www.kew.org/wakehurst/whats-at-wakehurst/millennium-seed-bank

OneZoom: Interactive map of evolutionary relationships among living organisms
https://www.onezoom.org/

Palaeontological Association
https://www.palass.org/

Svalbard Global Seed Vault
https://www.seedvault.no/

Wollemi Pine Conservation Program, Royal Botanic Garden Sydney
https://www.rbgsyd.nsw.gov.au/science/our-work-discoveries/germplasm-conservation-horticulture/wollemi-pine-conservation-program

World List of Cycads
https://cycadlist.org/